Smart Cities

Stan McClellan · Jesus A. Jimenez
George Koutitas
Editors

Smart Cities

Applications, Technologies, Standards, and Driving Factors

Springer

Editors
Stan McClellan
Texas State University
San Marcos, TX
USA

Jesus A. Jimenez
Texas State University
San Marcos, TX
USA

George Koutitas
Gridmates Inc.
Austin
USA

and

Texas State University
San Marcos, TX
USA

ISBN 978-3-319-86612-3 ISBN 978-3-319-59381-4 (eBook)
DOI 10.1007/978-3-319-59381-4

Softcover reprint of the hardcover 1st edition 2017

Printed on acid-free paper

This Springer imprint is published by Springer Nature
The registered company is Springer International Publishing AG
The registered company address is: Gewerbestrasse 11, 6330 Cham, Switzerland

Contents

Abstract

This book reviews the applications, technologies, standards, and other issues related to "Smart Cities." The book is divided into broad topical sections including Vision & Reality, Technology & Architecture, Transportation Considerations, and Infrastructure & Environment. In these sections, the authors who are experts in their fields, presents essential aspects of applications, technologies, requirements, and best practices. In all these cases, the authors have direct, substantive experience with the subject, and present an important viewpoint driven by industry or governmental interests. The authors have participated in the development and/or deployment of constituent technologies, standards, applications, and share unique perspectives on key areas of the Smart City.

Part I
Vision & Reality

Chapter 1
Smart Cities: Vision on-the-Ground

Ted Lehr

Introduction

“Smart City” is a puzzling expression. Every day, cities around the world are hearing and reading about ideas and projects that are titled and initiated under the word “smart.” Many of these ideas and projects are technology-laden. Mention of green energy, artificial intelligence, Internet-of-things (or otherwise connected “things”), self-driving vehicles and more is common. All of them or nearly all of them have voracious data demands.

So is a Smart City one that deploys cool technology to produce, ingest, and analyze data, then connects the parts and replaces infrastructure and services that do not do that? Is a Smart City a city that only uses twenty-first-century technologies? No. Rather, a Smart City explores, experiments with and uses these technologies to improve its community. The resulting improvements produce better governance, services, economic and education opportunities, and social equity for community members. It is almost certain that a city is not yet “smart” even after it has deployed and connected all the new technologies if it has neglected to include new, innovative policies that “understand” the nuances of privacy concerns and other issues in a world of ubiquitous data, social media, and high-performance computation and analytics.

As a data architect for the City of Austin, Texas (a photograph of its downtown is shown in Fig. 1.1), I am most familiar with this fast-growing metropolitan city’s efforts in understanding, defining, and creating smartness. In 2016, Austin was a finalist in a US Department of Transportation (USDoT) competition called “The Smart City Challenge.” Although Columbus, Ohio eventually won the USDoT

T. Lehr (✉)
City of Austin, Austin, TX 78704, USA
e-mail: Ted.Lehr@austintexas.gov

S. McClellan et al. (eds.), *Smart Cities*, DOI 10.1007/978-3-319-59381-4_1

Fig. 1.1 Austin, Texas (Ed Schipel—Flickr)

competition, the leading investment made by the City of Austin in engaging the surrounding community, its universities, and its industry partners brought together an understanding of what Austin thinks smartness is.

Vision

Austin's Mayor Steve Adler described Austin's vision for a Smart City [1]

> We could focus on what I call the shiny pennies, such as the futuristic transit stations, the automated cars, and the traffic lights that automatically adjust to weather and congestion. They're even talking about apps that can tell you where open parking spaces are so you don't have to drive around looking for them. We're talking about really cool stuff here, but that's not how our final application is going to read.
>
> [Instead], we will talk about how our top priority is making sure people can access work, school, and healthcare. Our proposal will begin with communities living on the formerly segregated east side of town called the Eastern Crescent. We will talk about people who were recently pushed out of Austin because it has become unaffordable, and how we will reintegrate them back into the flow of the city with more mobility options. The test for our success … will not be whether we can design the most sparkling technological toys. The test will be whether a senior citizen of the Eastern Crescent can get to her doctor without having to take a bus two hours in each direction.
>
> It is only by providing equal access that we will ever get on equal footing. We don't have a shot at racial equality without everyone equally and equitably benefiting from the services the city offers. The Smart City Challenge is not about turning Austin into an

ever-more-perfect utopia. It's about becoming an ever-more-equal city by creating opportunities and greater mobility for the people who are often last in line for the next big thing. This is less about transportation than it is about transformation.

So in essence, a Smart City is a city that uses technology and policies to improve its community. How that improvement manifests itself in the deployment and management of technology (assets) or local policy (process) is a topic of intense local interest, which may be influenced by standards, best-practices, and socio-economic value judgments.

The Role of the Private Sector, Universities, and Nonprofits

One stereotype of government in action might be that long procurement processes, caution, risk aversion, inefficiency, slow, low or sloppy technology pervade its operations. Aside from how comically disagreeable such a stereotype might be, it also belies and insufficiently respects the systemic complexity of a city's social, environmental, and fiduciary responsibilities. These responsibilities include the impact of the necessity of public oversight of the city's operations, the justifiable exclusion of most business models that generate profits in the private sector, and the abiding, natural constraint that cities do not have the option of picking their customer demographics and laser-focusing their "products" on a targeted set of customers to maximize revenue.

How does a city get "smart" under these constraints? Extending the constituency of a Smart City beyond its government entities to include public–private sector partnerships, nonprofits, universities and the community's open government enthusiasts will enable useful outcomes. By including this larger constituency, we can simultaneously innovate faster, leverage more advanced and nimble risk management processes, benefit from private sector and university expertise, and build a community of "smart" collaboration.

A Smart City engages this expanded constituency to explore innovative business and financing models for projects and programs, to explore advanced data analytics, such as sharing, mining, and integration technologies that will fund and jump start the use of data in business and city services, and to develop new data technologies that are sustainable, scalable, and valuable.

Private Sector

The private sector, including multinational corporations and local businesses, is usually more nimble in responding to change and more proactive in investigating and adopting new technologies. Additionally, a for-profit entity brings an ROI focus to its operations that the market place demands. Smart City governments need to explore innovative ways to engage the private sector to leverage these strengths. Through careful adoption of certain private sector characteristics, smart cities may be able to improve their communities and the services cities provide to the citizens.

Universities

Universities are sources of high-quality research. In most cases, these research problems and solutions are outside the pressing business interests of the private sector. Communities should work with universities to construct research projects and programs

- To understand complex systems, including the data that feeds into and are generated out of those systems;
- To understand what needs to be measured in order to *understand* those systems;
- To understand what needs to be measured to *improve* those systems, and avoid unintended consequences; and
- To provide a basis for new city policies and solution service models.

Universities also have extensive experience applying for and attracting grant funding. Many cities would benefit from their assistance in a coordinated approach to pursue business venture funding or research funding for projects that benefit both scholarship and community.

Nonprofits

A city's nonprofit entities are often "close to the pulse" of most of a community's activities. Including nonprofits in the definition and implementation of smartness helps ensure that important, inclusive policies and priorities are part of the discussion. Nonprofits are also sources of useful and even provocative data sets.

Smart Technologies: Generating and Consuming Data

Whatever technologies a city chooses in its drive toward "smartness," the technologies will have at least one characteristic in common: they will consume and generate large amounts of data. The rest of this chapter will discuss the challenges facing cities in choosing technologies, business models, and priorities for smart systems. The role of data will infuse and impact that discussion.

The role of data will also be one of the themes for the rest of this book. The reader will note that no system claimed as smart will be a laggard in the generation and consumption of data.

Austin, Texas: A Smart City

Before discussing some smart Austin challenges in detail, it will be helpful to describe the environment in which they are being considered.

Austin, Texas is one of the fastest growing major cities in the United States, and it is surrounded by communities which are among the fastest growing in their population categories [2–4]. The population of the city of Austin has been growing in excess of 100 residents per day since 2010. It has twice as many residents as it did 30 years ago, and its urban metropolis includes parts of five counties with approximately 2 million residents. Commensurate with the growing population is a robust economy, one that actually gained jobs during the recession that began in 2008. Much of that economic growth is driven by technology innovation, a vibrant startup scene, and creative industries that have established Austin as a destination for art, music, film, food, media, and culture. A commercial, housing, and revitalization boom has accompanied the population growth, and has added complexity to the city's composition.

The prosperity represented and derived from the ongoing population growth is not being shared equally by everyone. In 2015, Austin was designated as the most economically segregated major metro area in America [5]. Traditional, long-standing communities of color in working-class neighborhoods are being dislodged and dispersed by gentrifying pressures and high housing costs. In their place, new boutiques and trendy hot spots abound. Lower priced Austin suburbs, meanwhile, are absorbing and re-creating concentrations of poverty in areas where there are few services to meet the needs of the underserved. Working families at all income levels are pressured to live farther away from their jobs in the urban core. Austin has a famously highly educated workforce [6, 7]. Even this generally affluent segment finds itself impacted by the scarcity of reasonably priced family housing near its increasingly urbanized workplaces.

Smart City Imperatives

The medusa-like headwinds of rapid growth, economic segregation, and widespread affordability challenges, conspire to plague Austin with some of the worst mobility challenges in America. They conspire to diminish the quality of life, culture, and community that Austin has known for decades. Augmenting the city with "Smart" technologies that do not deliberately contribute to solutions to these challenges, or that do not create real and unforseen opportunities are "smart technologies" only in name.

In Austin, smart programs

- Must measurably deliver on the important community values of equity, economic, education and health opportunity, affordability, and environmental stewardship.
- Must provide testable hypotheses and pilots for repeatable, scalable adoption by other communities.
- Must build a foundation for Texas-wide partnerships on Smart City policies, services and outcomes.

In the absence of these or similar imperatives, Smart Austin will be just a denigrating label connoting a playground for technologists.

Three Dimensions of Smart: Projects, Policies, and Language

Austin's Smart City vision continues to change. In fact, even as this chapter is being written, Austin city staff, elected officials, community representatives, private sector leaders, and university researchers are having conversations that are at once spirited, innovative, methodical, frustrating, serious, playful and hopeful about the challenges and imperatives of "smart cities."

The rest of this chapter discusses three examples of Smart City challenges. These examples are but illustrations of the kinds of challenges and opportunities a city faces. Deeper and broader discussions can be found in the rest of the book. The example challenges are

1. What is a smart project?
2. What kinds of policy changes encourage smart activities?
3. What is the language of smart engagements?

Smart Projects

The following projects are derived from Austin's US Department of Transportation Smart City Challenge Proposal [8] (Fig. 1.2).

Austin proposed "Smart Integrated Mobility Solutions" consisting of the projects like the following:

Fig. 1.2 Integrated smart Austin mobility solutions (Austin Transportation Department)

- **Smart Stations** bring together a wide variety of mobility, health, retail, and other services and opportunities useful to travelers. One common characteristic of contemporary American cities is the need to travel "all over town" for work, education, food, medical, and other services. One might imagine a twenty-first-century implementation of traditional or even "old world" transportation hubs that not only take travelers to work and home, but have medical, legal, and grocery services that emphasize convenience and equitable access. These stations would also serve as centers for the deployment of autonomous and connected vehicles, urban freight logistics, and electric fleets (Fig. 1.3).
- **Connected Corridors** that link the Smart Stations, with new transit services (including a deployment of electric bus rapid transit), dedicated transit lanes and signal priority, better facilities for active transportation, and a sensor and beacon-rich environment that allows not only deployment of V2V and V2I connected-vehicle technology, but also encourages exploration of how to use the connectivity to improve community cohesion and prosperity.
- A **Mobility Marketplace** that connects travelers to their best packaged mobility options and provides an ecosystem for the development of new mobility services, with integrated payment options (including options for the unbanked) real-time travel information via app or kiosk. To ensure that a technology-based marketplace is accessible across the digital divide, this would also include a human-driven outreach component—Smart Ambassadors who will work on the neighborhood level to educate and help people take advantage of the marketplace.

Fig. 1.3 Smart stations (Austin Transportation Department)

All three of the above pilots are integrated into a set of **Ladders of Opportunity Initiatives** [9] that use Smart Stations, Connected Corridors, and the Mobility Marketplace to improve access to jobs, education, healthcare, healthy food, and other areas of need.

Supporting these kinds of programs require infrastructure and technologies that include

- A "One System" regional operations and management concept, integrating and enhancing travel management operations between the City of Austin, Capital Metro, Central Texas Regional Mobility Authority (a toll road operator), the Texas Department of Transportation, and other communities.
- A network of rich intelligent sensors that will feed more and better data to transportation agencies and help tackle a variety of persistent challenges facing local operations managers, public safety agencies, and planners.
- A two-way open data portal, known as the Data Rodeo, which will integrate and curate data from public, private, and nonprofit sources, including data providers beyond the transportation sector, and make data available to enable research and education as well as support application and tool developers.
- An Urban Analytics and Policy Research platform that plays an integrated role in performance management, metrics, and evaluation for the entire Smart City effort. All of these efforts would be supported by regular stakeholder engagement with rapid assessment and documentation of both successes and failures,

guided by a governance structure that brings together local agencies and non-profit partners in a consortium model, with executive-level authority residing within the city organization.

Smart Policies: A Smart Kiosk Example

Smart Kiosks [10] are increasingly popular technology items being proposed by vendors and others for providing interactive community portals to way-finding, search engines, browsing, Wi-Fi, advertising and social media. Physically, they are often about the size of a tall twentieth-century phone booth, with video screens and contemporary styling.

These kiosks are often proposed for high pedestrian traffic locations or community centers. In keeping with Austin's interest in using smart technology to strengthen and improve communities, these technology stations are being considered for areas in Austin wishes to develop into community centers. There is an interest in piloting various implementations since Austin wants to explore the positive and negative uses of the kiosks so that we iteratively improve on their uses.

Cities implement and abide by laws, even if those laws do not have language concepts for smart technologies. For example, Austin's sign ordinance was originally crafted in the 1980s. Although it has been widely praised for its control of blight in Austin, it was written before the Internet, smart phones, Wi-Fi, web browsers, search engines, and way-finding maps existed. As such, smart technologies with visual components, such as smart kiosks, often find themselves reviewed under admittedly outdated rules.

For example, is a kiosk that provides search directions to nearby businesses (as well as their websites) just another kind of advertising sign or is it something very different and unknowable by ordinances conceived in the 1980s? Amending those ordinances to understand and regulate smart technologies is a long community discussion that necessarily includes diverse and often incompatible viewpoints.

To inform those discussions, some cities are implementing temporary exemptions from ordinances in carefully selected geographies or "corridors" so that the communities can explore uses of smart technologies and the reaction of residents and businesses to them.

Smart Language: Assets, Valuations, Cost and Projects

Let us imagine that you are part of private sector, university or non-government organization and you have worked for several months helping your community and city imagine, design, plan, and commence implementation of a great Smart City project. Suddenly, near the end of this process, you learn that no one budgeted for archiving all the data the project is generating.

> "We thought the data was for operational uses".
>
> "We are only keeping a standard, fixed, multi-month revolving window of data to comply with open government rules. The City will delete data once it ages beyond that window".

The possibility of such a scenario arising is mitigated if you and the community understand the language and concepts used by cities. That language is an under-analyzed and often ignored complication in public–private partnerships: the language of the concepts, motivations, laws and rewards guiding government, public sector and university might be the same, but the meanings are often subtly different.

The likelihood of success of a community's smart projects increases if the meaning of language used is commonly understood.

In scenario above, data was not considered an asset. Data is not something physical. It cannot be "touched" or pointed at. For many cities, assets are physical. You can point at them. You can touch them. If data is not an asset, it cannot be assigned a cost or a value.

How do we talk about data so that a city can design and provide for sustainable archival, analysis and use of the data generated by a city's smart systems? There are four basic steps. They involve understanding how cities define and use concepts like assets, valuations, costs, and projects. If you do not express the thing you care about in these terms, you increase the risk that the organism we call a "city" does not even detect its existence.

The four basic steps for data are

1. Make data an asset.
2. Assign a cost to the asset.
3. Designate and measure the value returned for paying that cost.
4. To an existing or planned project, attach the asset, its cost and metrics for assessing the return value.

Let us consider each of these in order.

Data as an Asset

One begins to make data exist in the language of a city by identifying the physical assets that generate, carry, use, or store it. Assets as known by cities, at least in the case of the City of Austin, are tangible things. Things you can touch. Signs, roads, computers, swing sets are all assets. Data? You cannot touch data; so it is difficult to assign it an asset code. The devices that generate or store the data, however, can be treated as assets. So sensors, networks, and storage are assets. Even non-touchable cloud storage can be an asset because it has physical equivalents.

Assigning Costs to the Data Assets

The costs of these data assets can be expressed in terms like

- Dollars per Gigabyte of storage.
- Dollars per Gigabit per second of network demand.
- Dollars per unit of data processing capacity.

These costs can be tiered. For example, raw data from sensors on a city's streets might have one cost. Data combined and curated to produce neighborhood or corridor specific data might have another, larger cost.

Designating and Measuring the Value Returned from the Asset

Why is the city going to spend public dollars on these data assets? What value will the community derive? Importantly, the valuation of the return on a city asset does not have to be and often should not be expressed only in financial terms. For example these are values

- Reducing pedestrian injuries at intersections.
- Increasing or maintaining neighborhood satisfaction with its local parks.
- Being able to measure, understand, describe and eventually control green-house gas emissions.
- Understanding whether there equitable use of city parks and services.

Although each of these values may have correlates in terms of dollars saved or generated, expressing them in human-centric terms uses the language that is understood by everyone and not just actuaries and accountants. It is best to describe the value in the context of stated city objectives like reducing traffic fatalities, increasing access to affordable health care, walkable communities, etc.

Finally, the valuations need to be measurable. When assigning a value to a data asset, define value metrics the city can monitor as well as target values so that the city can assess the return it is getting. For example, base lining and tracking pedestrian traffic injuries at a set of intersections with a goal of reducing them by 30% would be a value metric and goal.

Attaching the Data Assets to Projects

Now that a city can recognize the existence, cost and value of the data asset, it cannot yet pay for it because the asset is not attached to any existing or planned and budgeted project. In the case of Smart City projects, there will be many data assets included. They need to be recognized as assets and assigned costs, values and metrics.

Do Not Forget to Consider the Data Market

Finally, you might ponder whether you want to propose that the city includes a cost recovery plan for the data assets. Instead of drawing on traditional revenue like the tax base, perhaps there are fees or other charges you can suggest the city imposes on external use of the data. For example, raw data might be free to the community, but curated or analyzed data might incur a charge. Or, if a business would like to make the data mission critical and therefore need Service-Level-Agreements on throughput and response time, you might propose a tiered pricing mechanism for these additional premium services.

Conclusion

When considering the definition and implementation of a "Smart City," the wise designer ensures that answers to the following questions make sense to the city leaders, the city inhabitants, and the constituents who partake of the city's services

- Do technologies such as sensors, beacons, and data make a city "smart"?
- Does the use of social media, way-finding, and other electronic or connected activities make a city "smart"?
- Is a "Smart" City energy efficient, automated, and environmentally friendly?
- Do citizens in a "Smart" City have access to justice, safety, work, education, culture, and affordable living opportunities?

This chapter has discussed a few of the challenges being managed by the city of Austin, Texas in its search for answers to these questions. In presenting these issues, we have provided a sample of challenges faced by the government, the residents, and the private sector partners in the development of a city intent on maintaining the values which continue to be relevant regardless of the technologies implemented. We believe this is the most accurate definition of a "Smart City."

References

1. Adler S (2016) Racing to the finish line in the Smart City Challenge. http://thehill.com/blogs/congress-blog/technology/276815-racing-to-the-finish-line-in-the-smart-city-challenge. Accessed 15 Mar 2017
2. Austin Planning and Zoning Department (2016) Demographic data. http://www.austintexas.gov/page/demographic-data. Accessed 12 Mar 2017

3. Austin Planning and Zoning Department (2016) Top ten demographic trends in Austin, Texas. http://www.austintexas.gov/page/top-ten-demographic-trends-austin-texas. Accessed 15 Mar 2017
4. Austin Business Journal (2016) Get ready for new neighbors: Austin to nearly double in size. http://www.bizjournals.com/austin/news/2016/10/11/get-ready-for-new-neighbors-austin-to-nearly.html. Accessed 20 Mar 2017
5. The Texas Tribune (2015) Study: Austin is most economically segregated metro area. https://www.texastribune.org/2015/02/23/austin-most-economically-segregated-metro-area. Accessed 20 Feb 2017
6. Austin Planning and Zoning Department (2012) Austin Metropolitan Area Education Attainment.http://www.austintexas.gov/sites/default/files/files/Planning/Demographics/MSA_ACS_2012_tracts_Educ_Att.pdf. Accessed 10 Feb 2017
7. Austin Chamber of Commerce (2015) The Central Texas workforce. https://www.austinchamber.com/economic-development/austin-profile/workforce. Accessed 15 Feb 2017
8. City of Austin (2016) Smart City Challenge. http://www.austintexas.gov/department/smart-city. Accessed 1 Feb 2017
9. US Department of Transportation (2016) Ladders of opportunity. https://www.transportation.gov/opportunity. Accessed 10 Feb 2017
10. City of Austin (2017) City of Austin wins smart city readiness challenge grant. https://www.austintexas.gov/news/city-austin-wins-smart-cities-readiness-challenge-grant. Accessed 20 Feb 2017

Chapter 2
Funding a Smart City: From Concept to Actuality

Stephen R. Galati

Introduction

Imagine a city of the future where citizens have access to all types of information and technologies "at their fingertips", from real-time bus schedules and traffic rerouting data to intelligent street lights and electrical energy systems. We have the technology to create these advanced "smart cities"; however, larger cities are plagued with large city issues, including increased urbanization and pollution. In fact, cities are highly urbanized areas that are responsible for the consumption of nearly 75% of the world's natural resources and the production of about 70% of the world's greenhouse gas emissions [1, 2]. Such environmentally unfriendly consumption and emissions is a threat to the sustainability of the city and the quality of life for future generations. Cities are in need of new solutions that tackle the harmful effects of urbanization while creating a citywide environment that is information-lush, environmentally sound, and attractive to residents and businesses alike.

Cities are starting to embrace the Smart City concept due in part to urbanization growth; an increase in energy and resource demands; a "smart" population with high-technology needs; and an infrastructure desperately in need of repair and renovation for future city loads. According to the United Nations, urbanization is growing at an alarming rate. In 1950, only 30% of the world's population lived in urban areas. By 2014, urban populations were at a sizable 54% of the global population. Analysts are now projecting this growth will continue, and that by 2050 more than 66% of the global population will live in urban areas. This projection means that, by 2050, an estimated 2.5 billion people will be introduced to the urban population [3].

S.R. Galati (✉)
TRC Companies, Inc., Palm Beach Gardens, FL 33418, USA
e-mail: SGalati@trcsolutions.com

S. McClellan et al. (eds.), *Smart Cities*, DOI 10.1007/978-3-319-59381-4_2

City governance is taking notice of urbanization trends as well as current projections for the Smart City market. Analysts estimate that the Smart City market will surpass the size of all traditional business segments and reach $3 trillion U.S. dollars just in the next few years [4]. Even more astonishing is the projection that cities around the world will invest an estimated $41 trillion by 2037 to upgrade citywide infrastructure and system connectivity [5]. City governance needs to stay savvy of their city's technology, mobility, and system needs to ensure their city remains a 'hot spot' to live, work, and travel.

The need for smart technologies and the Smart City model is burgeoning. The task of envisioning a successful and sustainable Smart City involves creativity and planning; however, the singular most important question in any Smart City design is: "How will it be funded?" Many larger cities seeking enhancement of their city service portfolios are forced to find alternate ways to fund their smart projects, particularly through public and private investors. Although most smart project investments are grounded in public funding [4], cities are also looking into the private and quasi-federal sectors for more funding options.

Investors, nonetheless, are hard to secure because investments into a Smart City are usually extensive and costly [4]. Additionally, investors, particularly private funders, want to know the probable return on their investment prior to committing funds to a smart project or program. Smart programs often require multiple funding streams to bear witness to positive results. Although the Smart City model is flourishing worldwide, little is certain about the ultimate significance of a smart project or how the intelligent technologies will truly add value to a city [4]. Planners and investors must have a shared vision of the Smart City and desired benefits to reap any type of program success. Ultimately, without project capital and the *right* committed funding, a Smart City vision will remain exactly that: just a vision.

The following chapter examines the complex process of identifying, pursuing, and securing funds for a Smart City program, and offers tools and considerations to help advance Smart City concepts from a vision to actuality. The chapter addresses common project considerations and potential funding sources for Smart City programs. Discussion about aligning potential funds with specific project components and the overall pacing of the smart project work follows. Finally, the chapter closes with the important matter of funding management. Without a fundamental understanding of project funding, Smart City governance will be hard pressed to succeed and may ultimately succumb to the seemingly indomitable challenges of planning and realizing a true Smart City.

Creating the Smart City Vision

Every great Smart City design begins with a vision of life enrichment for those who live, work, visit, or lead the city. A Smart City vision is relatively simple in concept; however, the process to realize this vision is far from simple. Creating a vibrant

Smart City concept involves many considerations ranging from project economics and feasibility to constraints and realistic execution. The following subsections outline important considerations to progress a Smart City vision of any complexity toward actuality.

Location, Location, Location

Successful smart cities have great locations. Chief among these considerations are locals that are primed for change and have enough outside appeal to validate a Smart City rejuvenation. The impact for such a citywide endeavor requires investment by the local agencies, a population ready for technological transformation, and an economic development landscape capable of sustaining revolutionary changes to the city. Creating the Smart City vision requires a location populous ready for the undertaking and a landscape ready for advancements.

Not all locations can make a Smart City vision work. Extremely rural areas or geographically vast areas are not ideal for Smart City solutions. Locations need some degree of population density and technological interests, along with the desired appeal to drive in citizens and travelers alike. With the right foundational basics, some lackluster locations can transform into enthralling smart cities.

What Are the Long-Term Visions of a Smart City Program?

Long-term visions for a Smart City differ from location to location, stakeholder to stakeholder. In less dense, less advanced locations, a Smart City may only involve advances in water quality and wastewater management, whereas larger urbanized cities in progressive nations will have much more robust Smart City needs and visions. Ironically, there is no definitive description of what constitutes a Smart City [6–8]. However, all smart cities involve newer technologies to solve existing quality-of-life issues. Long-term visions of the Smart City encompass quality-of-life improvements for citizens while employing intelligent technologies to enhance the appeal of the area.

Many Smart City models incorporate integration of multiple systems through a network of information and communication technologies [9]. Other long-term visions of smart cities are grandiose, involving a mix of intelligent transportation systems, "green" infrastructures, seamless connectivity, economic development, and citizen social improvements. Incidentally, with populations becoming more connected Smart City visions will undoubtedly incorporate intelligent transportation systems. Think of using a mobile phone to get real-time estimates on train and bus arrivals all while staying connected to your fare card balance and geolocation of your vehicle, and while riding on smart roadways with LED stoplights and pollution-free electric vehicles. Intelligent transportation systems will play a larger role in Smart City mobility, commuter data integration, and multimodal transportation operations within cities, towns, and communities [10]. Ultimately, intelligent transportation systems and projects aimed at enhancing the city's quality of life are principal elements to most Smart City conceptual visions.

Who Are the Stakeholders?

Smart Cities are generally located in medium to large urbanized areas and contain stakeholders interested in advancing the area for both economic and holistic benefits. Smart cities encompass the latest technologies aside a foundation of intellectual resources and developed infrastructure [11]. Citizens with the Smart City architecture are technologically savvy, whereas the governance with the city has interests in maximizing user data, connectivity, and intelligent technologies. Stakeholders are often varied and, depending on the size and geographic locale of the smart program, stakeholders can range from residents and private business owners to city economic development corporations and the government. Smart city stakeholders generally comprise local governments, research institutions, grassroots movements, technology vendors, business owners, tourism boards, and property developers, all of whom hold sometimes conflicting interests in designing a Smart City [6].

In heavily urbanized areas, the citizens can be the most powerful stakeholders in a Smart City and, as such, need to be involved with "smart" planning. These "smart" citizens need to be trained by local governmental agencies to not only be able to embrace the new technologies surfacing in their city but also serve as invested stakeholders in their growing city. Additionally, having smart citizens, in tandem with Intelligent Transportation Systems, smart technologies, and citywide interconnectivity, is crucial to a sustainable and healthy Smart City [12].

Stakeholders have very distinct desires for investing in a Smart City, including economic development, advancing technology use, and personal capitalism. When smart projects are surfacing, individual stakeholders are aligning with other individual stakeholders to develop mutually beneficial and prosperous smart destinations. These partnered stakeholders are looking beyond citizens and city services to a more dynamic return on investment through tourism [13]. Partnered stakeholders can collectively translate their differing interests into a broader tourism pull, and generate new income streams to the Smart City. Stakeholders, if deeply invested, possess the power to transform a technology mecca into a desired travel and business destination.

Understanding "Lighthouse" Projects

Not every urbanized area or city has the right components to become a Smart City. Without invested stakeholders, a pool of potential infrastructure necessities and a population ready for technological change, a city is unlikely to adequately sustain the Smart City architectural model. Then there are those urbanized areas that are destined for the Smart City model. These areas have a colloquial name: "lighthouse" cities. The term "lighthouse city" refers to the convergence of city governance, administration, bureaucracy, and city planning to form the groundwork of a Smart City model [14]. The imagery of this convergence of city leadership can be likened to the converged beam of a lighthouse, whereby city leadership, like the beam of light, progress in a singular, unified direction. Projects within lighthouse cities may be called "lighthouse" projects. Individual Smart City projects are considered "lighthouse" projects, with both terms used synonymously.

Smart City Project Considerations

Smart city programs are complex, diverse endeavors that encompass various new technologies, sustainable designs, and humanistic innovations. Smart city project considerations include, but are not limited to:

- New technologies;
- Intelligent transportation;
- Energy usage and efficiency;
- Digital automation;
- Public security and resiliency;
- Sustainable processes and urbanization;
- Hybrid approaches to manufacturing;
- Connected residents, workers, and visitors;
- Innovation;
- Economic impacts; and
- Return on investment.

These considerations have a common thread—centralization. Perhaps the greatest consideration aside from the above specific program concerns is how these individual components are integrated with one another, and how the Smart City governance can use generated information from the components to enhance the ease and quality of citizen lives.

From personal experience, intelligent transportation systems offer a robust, integrated mechanism for city resident ease, and simultaneously impacts many other aspects in Smart City living. Intelligent transportation systems lead to improved road and pedestrian safety; better management of and user insights into traffic; heightened vehicle connectivity; and enriched driver comfort [15]. This "smart mobility" consideration offers a blending of new technology, planned transportation hubs, and ease of population movement. Additionally, intelligent transportation solutions are fantastic assets to Smart City programs because they have sound architecture, optimized technology, economic sustainability, and are environmentally friendly. However, if a smart transportation solution is not the right fit for a specific Smart City visionary, it should not be forced as a consideration. With any project component, supportable solutions are ones that are encouraged by the citizens and backed by city governance. Without ridership and user interests at the core of a transportation design, any funded intelligent transportation or smart mobility system will fail and eventually lose funding [16]. City planners must remain creative and visionary, yet realistic, with their overall smart program concept to provide a solid path that leads to success.

Planning a Realistic Path to Reach the Vision

Smart city planners must develop a holistic yet realistic path to move from the concepts of their Smart City architecture to triumphantly achieving the program objectives. The initial step beyond conceptualizing the Smart City is discerning the necessary smart projects within your program and researching all potential funding sources to realize those projects. Since projects are complex and multifarious in design, Smart City planners can match particular elements of a project or sub-parts of a program with accessible funding sources. Part of this funding exercise is to prioritize projects from the "must do" projects to the "would be nice" projects, and put concerted effort for those smart projects with the greatest benefit to the city, the citizens, and to society as a whole. Additionally, when planning out the path to reach the overarching vision of the Smart City, planners need to recognize that delays occur on projects, that there may be unexpected obstacles, and that all funds used on smart projects must be carefully managed. Without meticulous care taken on every step of this pathway, project execution blockades may unceremoniously surface while the sought Smart City vision may remain only a vision.

Funding Sources: Lighthouse Projects and Smart City Programs

Any Smart City endeavor, whether it is for a small "lighthouse" project or a large Smart City program, is an expensive process. From conceptualization and planning to development and implementation, costs rapidly rise and can easily cause a much needed project to perish. A key step in any Smart City effort is to evaluate all potential funding options early and determine what financing source, or what combination thereof, offers the best fiscal solution to bring the project from idea to realization.

Figure 2.1 portrays the most common funding sources for "lighthouse" and Smart City endeavors. The mix of financiers extends from governmental and public entities to private and creative financing pathways. As garnered from the array of funding sources, Smart City projects have piqued the interests and support from various levels of government, stakeholders, and residents. The fiscal sources

Fig. 2.1 The potential funding options for Smart City projects are numerous and can be combined to form a collective, program-wide funding solution

described in following sections have varying levels of availability and particulars, along with distinct benefits to the development and long-term sustainability of the smart technologies employed.

Government-Level Funding

For many Smart City programs, government-level funding is a critical component to whether the intended project can be completed as originally envisioned and proposed. Government funding comes from country-sponsored (or federal) agencies, and is generally one of the first funding options considered by Smart City developers. Governments are generally well invested in the urbanization of areas. They use the availability of their capital funds to remain connected to the holistic urbanization, and ensure a measured increase in the country's quality of life.

Government-level funding is beneficial to both Smart City developers and to the Government. This form of funding allows transparency to and from the citizens, and enables governmental access to different types of user data. For example, in a Smart City, governments can obtain critical data and metrics related to the use and efficiency of their transportation systems. Access to this data allows Governments a way to improve their service platforms and heighten their electronic governance (or e-governance) of citizens. In fact, e-governance is a significant component to intelligent populations (smart cities) by allowing public agencies a viable method to uphold transparent, proficient, and expeditious administrative services [17]. Furthermore, government-level funding is a straight forward and practical approach to secure capital for a smart project. Funding may or may not have to be repaid over time; however, the mutual economic and user information benefits to the Smart City and Government may serve as a quick return on the investment.

Local-Level Funding

Smart projects are targeted to large urbanized areas and smaller improvement zone communities alike. A powerful funding source for targeted smart projects comes from local-level funding sources. Incidentally, most investments made in a Smart City program occur through public funding sources [4]. Such local-level sources include public development agencies, local economic development corporations, city/state/providence sources, and other locally invested quasi-agency organizations, such as utilities. Local financiers are generally highly invested in the urbanized area and may reap the rewards associated with Smart City technologies and projects. Often local financiers designate monies for local region development with the hopes of driving population, the local economy, tourism, and the area's attraction to new businesses, Local-level funders are stakeholders looking at a return on investment and may require repayment over time through tax dollars or land allotments.

Community-Focused Funding

Another source of smart project funding comes from community-based sources, such as grass roots environmental community groups, large businesses invested in a community, local businesses looking for area rejuvenation, and targeted project economic stimulus. Unlike the local-level funding sources that are invested in large to small urban areas, community-based funding sources target individual communities within the urbanized or improvement areas. These community-focused funders usually have a vested interest in aspects of a larger smart project or interest in singular smart projects that can benefit a specific community. Often, community-focused funding pays for part of a larger Smart City project and not whole programs. This form of funding is a good way to supplement the costs of a total project and should be used in tandem with other funding sources. Community funds can be considered, for the most part, an approach to match other types of source funding.

Public–Private Partnerships (PPPs)

Many smart projects have significant benefits for both the private and public sectors, and can generate increased consumer mobility and economic gains. For these reasons, Smart City projects may be funded through Public–Private Partnerships. A Public–Private Partnership, commonly known as a PPP or P3, is an "arrangement

(s) between government and private sector entities for the purpose of providing public infrastructure, community facilities and related services" [18]. PPPs are desirable funding mechanisms for smart projects not only for the mutual economic benefits but also for the sharing of capital investments, program risks, oversight, and responsibilities among the partners. Additionally, with multiple partners invested in a smart project or program, the likelihood of successful completion of all phases increases due to the additional project oversight and union of financers wholly invested for a successful outcome and quick return on investment (ROI).

Loans and Municipal Bonds

Oftentimes funding sources are not available, hard to acquire in a reasonable timeframe that aligns with the project timetable, or inadequate to fund an entire smart project. Developers can turn to more traditional sources of funding, including loans and municipal bonds, to help supplement other project funding or keep smart projects aligned to the project schedule.

The upside to using loans and "muni" bonds is, after obtaining the appropriate approvals, the lump sum capital made available to pay for the project work. Monies from loans and bonds can be used as fast or as slow as necessary to pay for the project. Smart projects can face delays due to a poorly timed distribution of funds or restrictions on how monies can be spent. Each funding source has its own peculiarities for releasing funds, which can negatively impact the project deliverables schedule. However, municipal bonds and loans make the funds available for spending when necessary and appropriate during a project. The lenders, that include investment banks, insurance firms, and the government, typically take on an oversight role, auditing project deliverables and ensuring the monies are being spent appropriately and as initially intended.

However, the downside to using loans and "muni" bonds is that all loans and municipal bonds must be paid back in full over a fixed timeframe. The duration is dependent on the financier's agreement and the specific loans and bonds used to finance the project, and often have set deadlines for payback, interest applied to the outstanding balances, and is constrained by amortization schedules. For the capacity of upfront project capital, project developers ultimately will spend more for the financed smart project than initially estimated by the end of the entire payback period. The loans and bonds funding route is a balance between funding availability and project cost returns, and is a decision left for developers.

Private Funding

Lighthouse or smart projects can be funded by private sources. Private funding is a viable option for both smaller projects with targeted stakeholders or aspects of larger projects. Rarely will private funding sources be used for entire Smart City programs or large, multifaceted, urbanized area projects. Private funders often have interests in specific aspects of a project including, among other reasons, economic development interests, tourism, better people mobility in congested areas, and sustainability. Private funding sources have constraints on project expenditures arranged by the financiers and agreed upon prior to releasing funds. Private funds often have expenditure limits per a certain fixed period (i.e., per year or per month). Smart projects using private funds are disposed to audits, oversight by a board of directors, milestone approvals for the release of funding, reduced or withdrawn funding, and funding inconsistencies. Private funders may also be susceptible to funding shortfalls, reduced or lackluster interest in a project over time, and impatience to any project delays. Overall, private funding can be a powerful approach to getting a lighthouse project completed; however, project developers must anticipate increased oversight, amplified "outsider" influence, and the need to reaffirm stakeholder interests.

User Charges and Pay for Performance

Smart city design and implementation is expensive regardless of location around the globe. New technology and advanced city transformations require upfront capital and a long-term economic plan to provide sufficient payback. Many Smart City programs are designed with a built-in repayment solution in the form of user charges and pay for performance. The idea is that users want an expanded and heightened city experience associated with a Smart City, and would be willing to pay a small surcharge for the experience. The concept is similar to the pay-for-use wireless networking ("WiFi") on an airplane, where users pay a small fee for the ability to stay connected at 30,000 ft.

Smart cities may choose to incorporate user charges into area taxes, city and utility bills, and other billable services. Some smart cities, such as in India, have begun applying parking fees, water and sewage surcharges, telecom fees, and utility (gas/electric power) surcharges to help pay for the available Smart City technologies [19]. Proposing these small pay for performance user charges helps the Smart City developer communicate a sustainable, fiscally sound vision for the Smart City concept. User charges are "built-in" funding sources that avail a stronger return on investment and a viable approach to paying back secured funds.

Smart City Challenges and Competitions

With the surge in technologies, city leaders often have lavish desires to transform their growing city into an urbanized, technological mecca—a place where people will want to live, work, and spend their money. Funding for a program aimed at meeting these lavish desires and endorsing a holistic approach to citywide mobility, accessibility, opportunity, connectivity, and sustainability can be difficult to secure. In fact, such a vision would likely require many funding sources collectively. However, these lavish desires for a transformed city are not impossible.

One of the most exciting demonstrations of Smart City funding has surfaced through organized and sponsored Smart City design competitions and challenges. In an effort to promote urbanized area qualities such as integration, mobility, economic development, and technological prowess, many countries have begun sponsoring Smart City competitions and conceptual challenges. These activities invigorate creative conceptual thinking and help stimulate progressive city projects. With funding difficult to secure for one-of-a-kind or progressive (and fiscally risky) projects, design competitions can bring the brightest futurists and conceptual thinkers together in a competitive setting to showcase how a holistic program or smart project can be realized along with the resultant quality of life and economic benefits. Ultimately, these Smart City competitions showcase ingenuity, problem solving, and fantastic ways to transform a city.

Here is just one example. In late 2015, the United States Department of Transportation launched a country-wide Smart City challenge focused on, of course, smart transportation. The challenge requested conceptual designs from mid-sized cities for a smart transportation system that could integrate multifarious data and applications throughout the city in tandem with the latest technologies. The overarching objective of the challenge was to help move people and goods throughout the city in an efficient, quick, and inexpensive way. Ultimately, after 78 applicants, and seven finalists, Columbus, Ohio was the winning city. As the winner of the challenge, the City of Columbus is now in the process of developing and implementing their *SmartColumbUS Vision* that boasts a connected transportation network, electric vehicle structures, and integrated data exchange. Through this competitive challenge, the City of Columbus gained the required funding and the one-of-a-kind opportunity to develop their Smart City program through a series of funded pilot projects [20].

Stay Creative and Vigilant for New Funding Sources

With the vast portfolio of funding sources just described, one might believe that every city can be a Smart City. This is hardly true. Although Smart City planners have many options, funding is still very limited and highly competitive. A magnificent Smart City design may not be enough to get it funded in whole or in part. Many factors play a role in Smart City decision-making including local politics, community opposition, and the inability to operate and maintain the proposed new systems. For example, high speed rail transportation thrives in Europe, China, and Japan; however, the United States consistently meets obstacles with planned systems related to gaining right-of-ways and the long-term operation and maintenance costs involved with these rail systems.

For new Smart City programs to be sustainable and successful, city planners need to remain vigilant and creative to find new funding sources. As stated before, a magnificent Smart City design is not enough to gain funding. Creative approaches to funding, including a matrix of funding sources, a progressive approach to matching funds, and an ironclad cost–benefit analysis, along with a solid design and invested stakeholders, can cue the magic needed to bring a Smart City concept to partial or full fruition.

Matching Project Elements with Accessible Funding Sources

An effective smart project funding approach involves a strategic plan to secure monies from different sources for different aspects of the whole project. Matching project elements with accessible funding sources offers significant advantages to any Smart City endeavor, but plays a vital role when more complex, larger scale Smart City programs need funding. For those large endeavors, single funding sources are rarely sufficient. As a result, a current trend is to move away from the large-scale Smart City planning approach and focus on small-scale integrated projects [4]. The following discussions briefly examine specific aspects of such a strategic and multifarious funding approach.

Specific Project Components

Designing a Smart City concept and eyeing the funding routes to actually build the Smart City involves a laser-sharp focus on the various project components. A Smart City is an amalgamation of four distinct design forces, namely urban features; knowledge and innovation economy; a push for technology push; and a pull for applications [18]. Understanding these forces on a smart design helps coagulate ideas for the fundable project components. Such components revolve around transportation systems, connectivity, new technologies, safety and security features, economic development, tourism hot-spots, citizen sociality, and sustainability, among other location-specific components.

Put in a simpler context, a prosperous, sustainable, and thriving Smart City requires a balance between economic, environmental, and social factors [6]. Focusing on these macroscopic areas of smart program interest allows easier discernment of important project components. For example, let us consider the local economics of a Smart City program. A significant consideration in the planning, funding, and design of smart cities is the effect on the local economy. Stakeholders want their "piece of the pie"; whereas governance and citizens want a significant level of economic competitiveness. Some economic concerns include job generation; area growth; diverse business products and markets; personal development, returns on investment, and enhanced productivity [21]. Consequently, specific smart project components can be presented systematically as the solutions for these economic concerns. This type of strategic thinking related to project presentation for fund acquisition is both creative and stimulating.

Sometimes these larger Smart City solutions are unwarranted and the better solution may be a smaller project piece. In large, urbanized areas, a Smart City solution needs to be dynamic, wide-ranging, and encompassing. However, in smaller urbanized areas, particularly in less progressive or poorer parts of the world, larger project components are overwhelming and not realistic. In fact, in less advanced countries, Smart City solutions are generally focused on specific local issues, such as better water quality and roads [22]. Smart city developers need to assess the local economy and determine realistic project thresholds for the smart project area. Additionally, project components must always be scaled for the particular local needs and always staying pure to the original Smart City concept—to help improve the quality of life for citizens in dense or urbanized areas [22].

Clustering Entire Program Elements

Once program elements are understood and determined, smart developers can straddle the power involved with clustering similar program components to determine new funding avenues and sources. Clustering elements, say under the banner of environmental or transportation, can avail stronger funding lines and visionary marketing messaging to secure particular monies. Having additional project components collected into a singular project can support greater returns on investment, dynamic integration, and even more striking quality-of-life benefits.

Clustering project components requires strategic thinking as well as a clear vision of program objectives and technology integration. With greater focus on how multiple parts of a program work together, Smart City visionaries develop a broader understanding of how to position the program to potential funding sources, how to capitalize on project component synergies, and how to inject efficiency into service overlaps. This sharper view of the program allows smart developers to recognize when to pursue monies for singular project solutions and when to pursue funding for a clustered program solution. This type of heightened understanding adds credence to the Smart City program's objectives and helps valid the overarching economic and society benefits for the program to be realized.

The Social, Environmental, and Economic Values

Regardless of whether smart project elements are clustered or kept singular or whether the targeted local is in the heart of an advanced region of the world or in a developing country, the core values of why a Smart City is desired must be upheld.

Smart cities must be built on an urban model to guarantee an increase in citizen life quality and social prospects, while sustaining balance with natural resources [23]. Put another way, a prosperous, sustainable, and thriving Smart City requires a balance between economic, environmental, and social values [6].

The social values of a society form the core incentives of a Smart City concept. As a city uses more intelligent technologies, the quality of living in that city should increase. Citizens want an easier way of life that heightens their ability to socialize and achieve work-life balance. Smart city solutions generally incorporate innovative and more efficient uses of infrastructure, which is ultimately needed to improve living standards and assuage the harsher living norms of economically weaker portions of society [1]. Other enhancements in society that are valued by citizens include greater connectivity to normal aspects of their life, such as transportation use, community events, and exciting new technologies. Funding and spearheading Smart City projects that enhance the social values of the city will produce increases in population, tourism, and economics.

On the flip side of this discussion, not everyone wants to live in an information-lush, technology-driven city. Citizens may want to unplug from the Smart City framework and go "ghost" for a period of time. These citizens would become disconnected from the hyper-connected society in which they live and rely on more traditional forms of connectivity, namely physical encounters and face-to-face communication. Such socially valued opportunities to become disconnected from the Smart City framework should be included in the overall design and strategy for the Smart City, thus allowing ample opportunity for human interaction and in-person social networking [24].

Working in tandem with a Smart City focus on society values is a dynamic view on sustainability and environmental values. Successful Smart City designers plan for climate change and incorporate new technologies into their concepts to minimize pollution, maximize sustainable solutions, and engage an environmentally safe approach to city life [1]. Citizen interests expand in smart cities due to, among other things, a citywide approach to environmental sanctity and "green" solutions. Cleaner living environments that are hyper-focused on maintaining future clean living is perhaps the common core to all environmental values related to a Smart City.

We have already spent a fair amount of time discussing the economic values and benefits of a Smart City program. However, a specific economic value surrounding Smart City projects involves governance and how the governing body can add layers of efficiency and cost savings into a Smart City. A major citizen appeal for the Smart City concept is the ability to have real-time city information; however, this large data processing is an enormous undertaking that requires significant technologies and resources. Data beyond a real-time time window is aged and has reduced value [25]. Economic value can be reached through the integration of available real-time data concentrated on very distinct, but important aspects of Smart City use. As efficiencies are developed in how governing bodies analyze and incorporate data, a wider array of economic value can be realized through a slow, but wider implementation of user data application. Collectively, these economic values, along with prominent social and environmental values, make Smart City solutions appealing to citizens, business owners, and governing bodies alike.

Prioritize Projects and Design Your Funding Approach Using an AFM: Accessible Funding Matrix

After funding sources have been researched and project determinations have been made, Smart City developers are left with hard decisions related to where they will put their efforts to gain funding and in what order. The activity of prioritizing the needed Smart City projects and designing the final funding business approach is often complex and requires a systematic methodology to ensure all major aspects of all projects are being considered. This exercise is in alignment with the availability of resources to pursue funds, the time necessary for proper Smart City messaging, and the realistic constraints of project-award pacing.

A particularly effective methodology to prioritize projects and develop the right funding approach is through an Accessible Funding Matrix (AFM). Figure 2.2 depicts a typical Accessible Funding Matrix. An AFM can be used for entire Smart City projects requiring funding or project scope components, such as Intelligent Transportation Systems, enabling technologies, sustainability features, and other project scope items. The AFM, similar in design to common risk matrices, serves as a template to recognize where on the spectrum a particular project lies with regard to overall society needs, benefits, and ease of funding. AFMs are particularly useful in understanding the needs of the Smart City governance who, with the help of new

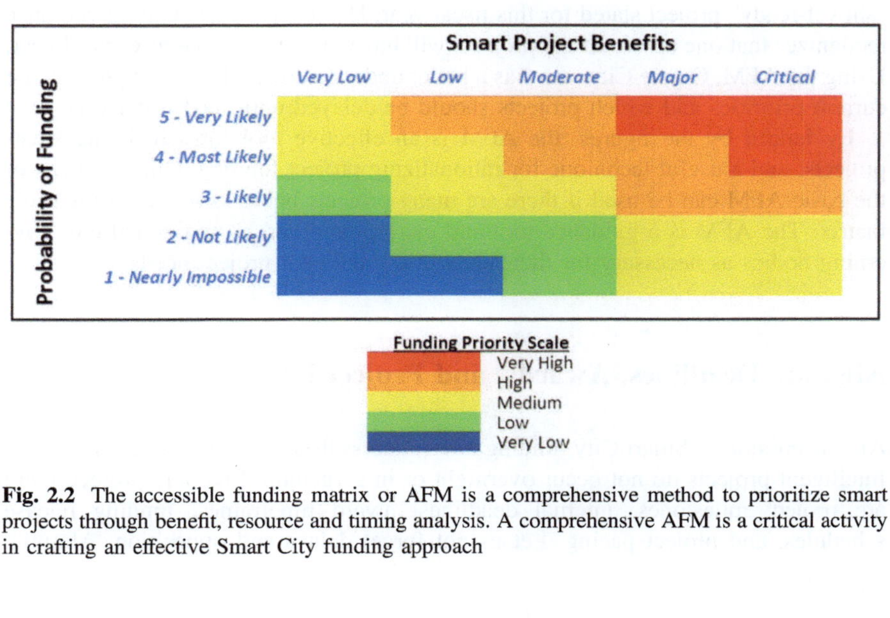

Fig. 2.2 The accessible funding matrix or AFM is a comprehensive method to prioritize smart projects through benefit, resource and timing analysis. A comprehensive AFM is a critical activity in crafting an effective Smart City funding approach

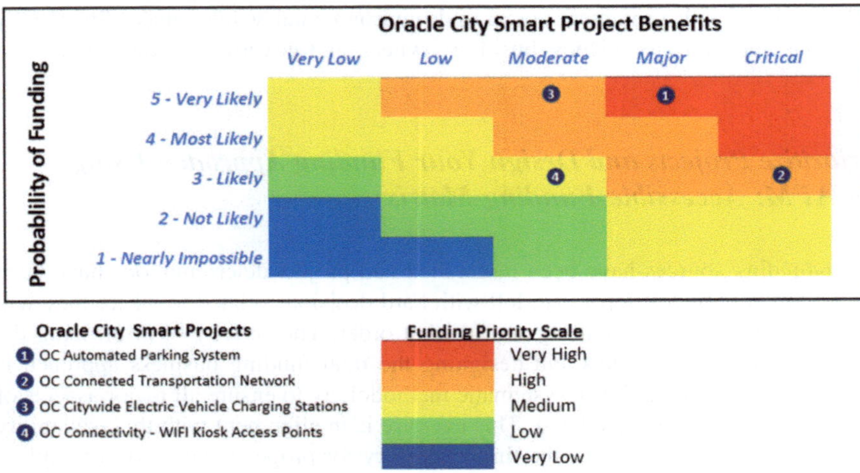

Fig. 2.3 An example Oracle City governance AFM which highlights four example Smart City technology projects: automated parking system; intelligent transportation system; citywide electric vehicle charging stations; and WiFi Kiosk access points

technologies and system data availability, are progressing from traditional delivery methods to more collaborative and integrated service models [26].

Figure 2.3 showcases an example Smart City governance AFM for a fictional Smart City: Oracle City.

In the example, the governing body at Oracle City has over 20 potential Smart City projects conceptually planned, but has limited resources for implementation. The City government used an AFM to determine priority for their four near "shovel-ready" project slated for this fiscal year. However, Oracle City governance recognizes that one or two of the projects will have to wait until the next fiscal year. Using the AFM, Oracle City now has a better understanding what smart projects are current priorities and which projects should be delayed until next fiscal year.

Evidenced by the figures, the AFM is an effective tool for scheduling smart projects, and a useful technique for rationalizing project funding. Other formats of the basic AFM can be used if there are many projects being assessed on the same matrix. The AFM is a guidance tool and can be modified by individual city governing bodies as necessary for their particular intelligent project needs.

Aligning Deadlines, Awards, and Project Pacing

Any discussion of Smart City funding must address the "after the award" timetable. Intelligent projects do not occur overnight or in a vacuum. Like any project, there are project milestones, internal deadlines, award timeframes, funding release schedules, and project pacing. Let us not forget delays and opposition. Aligning deadlines, awards, and project pacing must occur during the planning phase prior to the pursuit of any funds. The following subsections cover appropriate elements of economic and project alignment with the overall objective to stabilize and secure a smart project.

Proposing with Project Pacing in Mind

Smart city projects are complex and are generally considered one-of-a-kind endeavors for the pursuant cities. Considering the mix of concept, funding, and realization, Smart City developers can often overlook the realities of project pacing. As example, implementing new equipment requires having funding in-place and available, while incorporating equipment manufacture and delivery logistics into the project schedule. Misalignment between these and other aspects of the project can cause terrible delays, result in lost funding, increased costs, and a domino effect with future schedule milestones. Without proper pacing and deliverable timing, a slippery slope of delays and shortfalls will surface among project scheduling and funding.

Aligning Project Scopes with a Realistic Funding Source

As mentioned earlier, Smart City programs may require a combination of funding sources to support the various project scope areas. In fact, Smart City plans that are realized involve the alignment and integration of intelligent project foci including economy, energy, mobility, community, and the environment [27]. Further, Smart City performance indicators encompass smart economics, smart people, smart governance, smart mobility, smart environment, and smart living [28]. Alignment among these indicators and project scope areas has significant advantages to lining up the right funding sources as well as the winning message to secure those funds. Without proper assessment and understanding of this alignment, Smart City developers will have an inefficient and ineffective funding approach that may ultimately shelve the particular smart project or critically hinder the conduct of the Smart City program. Clean alignment among project factors and indicators feed into another significant step in Smart City realization: the cost–benefit analysis.

Cost–Benefit Analysis: Definition and Importance to the Smart Project

A critical aspect to the long-term success and positive ROI of a funded smart project, regardless of size or complexity, is to conduct a Cost–Benefit Analysis (used synonymously with Benefit–Cost Analysis). A Cost–Benefit Analysis (CBA) holds different meaning for different governing bodies and stakeholders. According to the Federal Government in OMB Circular A-94, a Benefit–Cost Analysis is "a systematic quantitative method of assessing the desirability of government projects or policies when it is important to take a long view of future effects and a broad view of possible side-effects" [29]. Another investor, the United States Department of Transportation, defines a Benefit–Cost Analysis differently, considering the analysis as a way to measure fiscal value of all the anticipated benefits and costs connected with all members of society [30]. Using this definition, the calculated benefits are linked with what all the people in society would be willing to pay to have the project built. If people are willing to pay more than the project actually costs, then the project will have positive net benefits (i.e., benefits minus actual costs).

Private stakeholders hold refined definitions of CBAs which are particular to business. As example, Heinzerling and Ackerman [31] from the Georgetown Environmental Law and Policy Institute suggested that a CBA sets a fiduciary standard for quantifying the success or failure related to government projects and programs. This definition highlights the significance of a CBA in project oversight and initial due diligence related to the feasibility of conducting a project. Ultimately, regardless of the definition applied, a CBA is used to better realize the reasonability and long-term sustainability of a project or program.

Further clarifying the need for this analysis, the CBA scrutinizes the prospective project from the standpoint of the citizenry, and accounts for the net benefits and net costs based on the criteria outlined by the funding sources. The analysis is conducted to answer the question, "Is society better off with the project or without the project?" The CBA addresses, at a minimum, travel time savings, operating and maintenance costs and savings, emission reduction, and economic development. Recognizing the overall benefit of the project, the analysis delivers a benefit-to-cost ratio for the project over a period of years (typically 30 years) and discounted to present value at alternative rates (generally 3 and 7% alternatives). The CBA provides a comprehensive overview of the project's fiscal performance throughout a long-term, fixed duration.

A CBA is a critical step in any smart project planning process and, if comprehensive in development, can offer durable proof-points to the legitimacy of the project. The analysis quantifies the overall benefits of a proposed smart project long before the "shovel hits the dirt." It enables short-term and long-term views of the proposed project with regard to the impacts and benefits to the surrounding community. Remember, Smart City projects are focused on improving quality of life within the area. The CBA offers a comprehensive understanding of why money should be allocated and spent on the project. Local businesses will rely on the CBA, as well, by enabling businesses and agencies to be forward-looking with regard to their long-term planning and revenue streams. Of greatest importance to local stakeholders is that the CBA addresses the benefits along all avenues of economic growth and public/community interests surrounding the project. Smart city designs embrace society needs and are concentrated exclusively on the citizens [22].

Ultimately, the CBA is a potent, forward-looking analysis of how the Smart City project will contribute to society and at what cost.

A powerful role of every CBA is to outline the fiscal feasibility of the project. Positive effects on society are great; however, the actual costs of the project will ultimately determine if the project is funded or not. Some common current cost discussions include, but are not limited to

(a) Realistic Short- and Long-Term Project Costs.
(b) State of Good Repair of Current Infrastructure.
(c) Current and Future Maintenance and Operational Costs.
(d) Costs related to Missing Infrastructure/Facility/Mobility/Connectivity/Technologies/Project Structures.
(e) Deepening Impacts and Needs of the Community (Monetized).
(f) Current Impediments that Cost Money, Wastes Time, and Lessens the Public's Interest or Value (Monetized).

Lastly, a CBA can serve as the "voice of reason" for why a lighthouse or Smart City project should be funded. Table 2.1 details some considerations when evaluating the required funding and lasting effects of intelligent projects. These considerations are important discussion topics for pitching smart project funding sources and showcasing sustainable economic and communal benefits.

Funding Management

Once project funding has been secured and the monies have been released to the Smart City officials, the task of funding management begins. Like any other project management, close fiscal oversight must occur. Smart city funders, whether public, private, or partnerships, want to ensure the money awarded is used appropriately and for the approved project or program.

The administrative task of funding management is not glorious or exciting; however, it is vital to maintaining a fiscal stream flowing into the project. Meticulous management of awarded funds is an absolutely critical step to safeguarding the likelihood of future scope components or smart projects within the same program will also be funded. This administrative task needs to be considered early on during the initial Smart City planning, and long before the conduct of the funding approach. Management of funds commonly involves documenting the appropriate use of awarded monies; compliance with the funding agreement; timesheet and invoice recordkeeping; comprehensive and timely reporting of expenditures; project milestones; conformance with regulated cost principles; and audit preparation.

One important component of expenditure documentation is the alignment of specific expenditures to distinct project deliverables. Often, and usually during an audit, funds have to be repaid to the funding sources because of a significant lack of documentation. In some cases, money awarded for one project is actually used on

Table 2.1 Considerations in funding or not funding a Smart City project or program

Considerations when not funding a smart project	Considerations when funding a smart project
Public safety concerns	Reduction of public safety issues
Environmental and sustainability concerns	Reduction of environmental concerns, such as pollution, emissions, greenhouse gases
Increased pollution/emissions	Addressing the public and community needs….keeping the public happy!
Increased public/community needs without resolution	Saving travel/commuting time
Increased maintenance and operational costs	Minimizing maintenance and operational costs
Increased public travel time and nuisances	New revenue streams and opportunities for additional capitalization
Increased public dissatisfaction	Maximizing public use and interest
Loss of revenue	New tourism opportunities, increased visitors, and more tax revenue
Loss of public use or interest	Fueling the local economy through economic development
Loss of tourism, visitor dollars, and tax revenue	Adding jobs in the area
Local economic downfall or recession	New infrastructure, technologies, connectivity and mobility
Other adverse community impacts (i.e., property value impacts)	Other positive community impacts (i.e., property value rise)
Think: How much will this "Non-Project" option cost over time!	*Think*: How much Money and how many Benefits will this "Project" option bring to the community over time!

another project under the same smart program. This is a form of misrepresentation and funding negligence, and can be used to penalize the Smart City developer or agency in charge of using the money. Clean, complete documentation of how every cent of the awarded funds is being spent is a necessary part of the entire Smart City funding cycle. Often, Smart City governance employs dedicated administrators to handle this documentation and provide oversight for the use of funds throughout the entire Smart City program.

Regarding audits, nearly all funders conduct periodic audits. Funding sources will usually ask for and should be given unconditional access to all documentation related to their funded project. Government sources typically have routine, scheduled audits and will require copies of timesheets, invoices, payment stubs, project milestone reports, technology testing results, and other bits of information. For example, the United States government employs what is called an A-133 auditing process, which necessitates (at a minimum) grant spending documentation compliant with federal administrative rules and cost principles, validation of allowable costs, memorandum of understanding and agreements (MOU/MOA), indirect cost itemization, and Buy American Act and Davis–Bacon Act compliance. Other countries have their own variation of auditing protocols that all Smart City governing bodies need to acknowledge, understand, and plan for. Orderly and satisfactory funding audits often lead to further Smart City funding. It is not uncommon for funding to be withdrawn and future funding pursuits halted after a poor audit.

The final discussion related to funding management encompasses repayment sources. Funding sources are keenly aware that Smart City governance will employ different ways to stimulate income. Earlier in this chapter, we discussed user charges and pay-for-performance fees. These forms of repayment and returns on investment must be disclosed during audits and tracked under the administrative task of funding management. Funding source will be interested to identify ways to get repaid over time through various smart technology charges and fees, particularly if the funded project has formed different income streams for the city. If repayment is part of the mutual funding agreement, the auditors will look carefully at the income streams to ensure that proper repayment is occurring, and that monies generated from the project are not being misappropriated or funneled to another Smart City activity. Ultimately, clean, comprehensive, and thorough funding and expenditure oversight will benefit the Smart City, the administration staff, the auditors, and future Smart City projects.

Conclusion

With global urbanization steadily rising, cities are turning to the Smart City model to help tackle the harmful effects of urbanization and offer society a citywide environment that is data rich, environmentally sound, and attractive to residents and businesses alike. The Smart City model was developed to improve the quality of life and economic opportunities for city populations. However, smart cities are not cheap and involve a strategic and comprehensive plan to find and secure the best funding for the individual smart projects or the whole program. These Smart City investors (funding sources) want to know the probable return on their investment prior to committing funds to the program. City governance, on the other hand, want to enhance their city's technology, mobility, and system capabilities to ensure their city remains a "hot spot" to live, work and travel. Ultimately, smart programs often require multiple funding streams to produce positive results for both investors and city governance alike, while delivering substantial quality of life and economic benefits.

The chapter described the intricate process of identifying, pursuing, and securing funds for a Smart City program. Potential funding sources include public and private investors, as well as some less customary sources such as municipal bonds, capital loans, and user pay-for-performance charges. The chapter presented tools and considerations that typically add value to Smart City approaches. The Accessible Funding Matrix, for example, offers considerable support to Smart City planners as they prioritize individual "lighthouse" projects and optimize their funding strategy. Solid funding management and project pacing are revealed as key ingredients to maximizing secured funds, minimizing smart program risks, and stabilizing future funding opportunities. A Smart City planner's end goal is always to secure the *right* funding for the *right* smart project to advance their Smart City concept from a creative vision to an extraordinary reality.

References

1. Hayat P (2016) Smart Cities: a global perspective. India Q 72(2):177–191. doi:10.1177/0974928416637930
2. The Global Commission on the Economy and Climate (2016) Retrieved from http://newclimateeconomy.report/2016/wp-content/uploads/sites/4/2014/08/NCE_2016Report.pdf . Accessed 3 Apr 2017
3. United Nations, Department of Economic and Social Affairs (2014) World prospects: the 2014 revision. https://esa.un.org/unpd/wup/Publications/Files/WUP2014-Report.pdf. Accessed 5 Apr 2017
4. Anthopoulos L, Fitsilis P, Ziozias C (2016) What is the source of smart city value? A business model analysis. Int J Electron Gov Res 12(2):56–76. doi:10.4018/IJEGR.2016040104
5. Smart America Challenge (2017) Smart cities USA. http://smartamerica.org/teams/smart-cities-usa/. Accessed 31 Mar 2017
6. Angelidou M (2014) Smart City policies: a spatial approach. Cities 41(1):S3–S11. doi:10.1016/j.cities.2014.06.007
7. Angelidou M (2015) Smart Cities: a conjuncture of four forces. Cities 47:95–106. doi:10.1016/j.cities.2015.05.004
8. Albino V, Berardi U, Dangelico RM (2015) Smart cities: definitions, dimensions, performance, and initiatives. J Urban Technol 22(1):3–21. doi:10.1080/10630732.2014.942092
9. Doran M, Daniel S (2014) Geomatics and smart city: a transversal contribution to the smart city development. Inf Policy 19:57–72. doi:10.3233/IP-140330
10. Egan S (2017) Smart cities and communities. Inst Transp Eng J 87(2):36–38. Retrieved from https://search.proquest.com/docview/1874710603?accountid=35812. Accessed 7 Mar 2017
11. Hajduk S (2016) The concept of a Smart City in urban management. Bus Manag Educ 14(1):34–49. doi:10.3846/bme.2016.319
12. Šiurytė A, Davidavičienė V (2016) An analysis of key factors in developing a Smart City. Bus XXI Century 8(2):254–262. doi:10.3846/mla.2015.900
13. Khomsi MR (2016) The Smart City ecosystem as an innovation model: lessons from Montreal. Technol Innov Manag Rev 6(11):26–31. Retrieved from http://timreview.ca/article/1032. Accessed 5 Mar 2017
14. Das G, Dubey R (2017) The smarter city: Jaipur is leading by example with its ahead-of-the-curve initiatives. Bus Today 26(1):46–50
15. Javed MA, Ben Hamida E, Znaidi W (2016) Security in intelligent transport systems for smart cities: from theory to practice. Sensors 16:879. doi:10.3390/s16060879
16. Cristian-Gabriel B (2015) Public transport approaches from a Smart City perspective. Young Econ J/Revista Tinerilor Economisti 12(24):67–72
17. Khatoun R, Zeadally S (2016) Smart cities: concepts, architectures, research opportunities. Commun ACM 59(8):46–57. doi:10.1145/2858789
18. Tahir MS (2017) Public private partnerships (PPP); innovations and improvements for future health care systems in Pakistan. Prof Med J 24(1):1–9. doi:10.17957/TPMJ/17.3833
19. Pratap KV (2017) Financing smart cities. Ministry of Urban Development, Government of India. http://smartcities.gov.in/upload/uploadfiles/files/Financing%20of%20Smart%20Cities.pdf. Accessed 12 Mar 2017

20. U.S. Department of Transportation (2017) Smart city challenge. https://www.transportation.gov/smartcity. Accessed 10 Mar 2017
21. Popescu GH (2015) The economic value of Smart City technology. Econ Managand Financ Mark 10(4):76–82
22. Bawany NZ, Shamsi JA (2015) Smart city architecture: vision and challenges. Int J Adv Comput Sci Appl 6(11):246–255. doi:10.14569/IJACSA.2015.061132
23. Keta M (2015) Smart city, smart administration and sustainable development. Rom Econ Bus Rev 10(3):43–56
24. Calzada I, Cobo C (2015) Unplugging: deconstructing the Smart City. J Urban Technol 22(1):23–43. doi:10.1080/10630732.2014.971535
25. Thompson EM (2016) What makes a city 'smart'? Int J Archit Comput 14(4):358–371. doi:10.1177/1478077116670744
26. Glasmeier A, Christopherson S (2015) Thinking about smart cities. Camb J Reg Econ Soc 8(1):3–12. doi:10.1093/cjres/rsu034
27. Mattoni B, Gugliermetti F, Bisegna F (2015) A multilevel method to assess and design the renovation and integration of Smart Cities. Sustain Cities Soc 15:105–119. doi:10.1016/j.scs.2014.12.002
28. Letaifa SB (2015) How to strategize smart cities: revealing the SMART model. J Bus Res 68(7):1414–1419. doi:10.1016/j.jbusres.2015.01.024
29. The White House, Office of Management and Budget (1992) Circular No. A-94 revised, subject: guidelines and discount rates for benefit–cost analysis of federal programs. https://obamawhitehouse.archives.gov/omb/circulars_a094. Accessed 9 Mar 2017
30. United States Department of Transportation (2012) Notice of funding availability for the Department of Transportation's National Infrastructure Investments under the full-year continuing appropriations, 2012; and request for comments. Fed Regist 77(20):4863–4880. https://www.gpo.gov/fdsys/pkg/FR-2012-01-31/pdf/2012-1996.pdf. Accessed 10 Mar 2017
31. Heinzerling L, Ackerman F (2002) Pricing the priceless: cost–benefit analysis of environmental protection. Georgetown Environmental Law and Policy Institute, Georgetown University Law Center. http://www.ase.tufts.edu/gdae/publications/C-B%20pamphlet%20final.pdf. Accessed 10 Mar 2017

Chapter 3
The System Complexities of Smart Cities and the Systems Approach for Standardization

Manyphay Viengkham

Evolution of the Element

In today's modern and increasingly connected society, technology continues to advance very rapidly, and with varying impacts across segments of industry and society. The continual increase in functionality produces an increase Life Cyclein data being generated, an increase in processing capabilities, and an increase in dependencies with other technologies all within smaller and more compact units. As a result, technologists, architects, and users are presented with the management and operation of increasingly complex systems.

An excellent example of this phenomenon is the humble and ubiquitous electric meter. When Thomas Edison formed the Edison Electric Light Company in 1878, he quickly needed a way to measure electricity consumption in order to bill his customers. Out of this need, he developed a chemical-based meter, which used two copper rods as electrodes in a chemical solution. The measured power consumption was based on the change in mass of the electrodes. Technically, the meter could only measure ampere-hours and did not have the functionality to accommodate any fluctuations in the voltage. As one can imagine, this meter reading method was challenging for the utility, inaccurate, and provided no means for the consumer to see their consumption until after the bill arrived!

In the early 1880s the pendulum meter was invented. The advance version of this type of meter used a counter, two pendulums, and coil that were connected to the voltage. This allowed utilities to measure the ampere-hour or watt-hours, but it could only be used with direct current power (DC). Unfortunately, the cost was also quite high for the utilities, primarily because the meter required two clocks. The pendulum concept meter was later replaced with a motorized wattmeter developed

---

M. Viengkham (✉)
IoT Center of Excellence, Microland Limited, Alpharetta, GA 30009, USA
e-mail: ManyphayV@microland.com

S. McClellan et al. (eds.), *Smart Cities*, DOI 10.1007/978-3-319-59381-4_3

in 1889 by Elihu Thomson at General Electric. This meter consisted of an iron-less motor commutator, with a rotor that operated based on the voltage running through a coil and resistor. Accuracy of measurement had improved compared to the pendulum and chemical meters, but again this meter was only designed for DC power, and the commutator was becoming an inhibitor because it generated very low torque and could only measure ten amperes or more.

A series of related inventions and discoveries resulted in the electric transformer, which allowed the industry to easily change voltages and thus expand into larger electrical systems. This technology produced a massive paradigm shift in the power industry, replacing DC transmission/distribution systems with the alternating current (AC) systems we are familiar with today. This transition also created a need for more precise and low-cost AC electrical meters, which was later filled by the development of induction meters. The discovery of rotating electric fields by Nikola Tesla in 1888 and their effect on the rotation of a solid disk [1] led to the development of induction meters, which then allowed for the corresponding development of AC meters in 1899. Following this was a series of improvements to measurement accuracy, reliability, manufacturing cost, and so forth.

By the turn of the century, the number of users connected to the nascent power grid had increased significantly. As a result, the need for various billing capabilities like prepaid metering, maximum demand meter, remote control meters, etc. also developed rapidly. In the 1960s and 1970s, the concept of remote metering was introduced, along with the incorporation of electronics into the meter. These advances played a significant role in the transition of a mechanical or component metering device into a complex system of machines, software, information, people and processes [2].

The Evolution of an Element to a Complex System

Today, electric meters are part of a larger network of systems which is often referred to as the "Smart Grid." The modern grid is driven by very sophisticated business and consumer requirements which include advanced reliability and resiliency as well as real-time information about outages, restoration time, and consumption. Additional requirements include flexible and advance pricing models that fit various consumer profiles as well as environmentally friendly and sustainable energy resources such as solar, wind, hydro, etc. All of these demands have molded and expanded the system as a whole, producing meters which are now very sophisticated and can interface with other devices, software, and systems. Today's electric meters capture important information that is crucial to the operation of the grid.

In the market today, the meter is part of larger system, often referred to as the Advanced Metering Infrastructure (AMI). AMI deployments include a meter capable of two-way communication, which is connected through a communication infrastructure to a network of other meters and data servers. The system as a whole enables the energy service providers to efficiently manage operational costs by reducing, or in some cases eliminating, the need for staff and field crews to manually read meters or address service issues. Detection of power theft is also a capability with the AMI system that was previously not available. Other cost savings related to AMI include improved outage management, which reduces the number and duration of outages. The AMI system also empowers the consumer to be more aware of their electrical consumption in real time. In addition, AMI provides billing flexibility, which leads to other special billing programs like pre-pay billing, Demand Response, and sophisticated energy efficiency programs. The result of all of these improvements is vastly improved customer service and a much more efficient power delivery system.

This explosion of capabilities is only possible with the advancement of the meter technology, computing systems, data exchange, and network technologies. If we break it down, the AMI system is comprised of several building blocks of hardware and software components. At the consumer level, the advanced meter records time-based data about power consumption sends events to the service provider which signal service interruption or anomalies in consumption patterns. The meter can also receive messages from the utility service center to disconnect power, reconnect power, or reset the reading. Information exchange from the utility service center to the meter at the home, and vice versa, is possible with network technologies like Broadband over Power Line (BPL), Power Line Communication (PLC), Fixed Radio Frequency (RF) and other public networks like cellular or fiber. In 2015 the EIA reported over 47 million advanced meters deployed in the United States [3]. These meters are generating data from less than one minute intervals to daily reads which can translate to gigabytes of data for certain size utilities.

Data from the meters in the field flows to the utility service center where it is typically received by an AMI software system providing data storage, data computation/analytics, and User Interface (UI) to allow the user to monitor and analyze the data and the network. This software also allows the user to send commands to the meter. The meter data can then be shared with the billing system to create a seamless billing process in addition to enabling billing higher accuracy and flexible billing options. Beyond the billing system, meter data is used by the Meter Data Management System for further consumption profiling and billing capabilities. The meter data can also be used by the Demand Response system, which enables the utilities to match consumption with supply. All of the system enhancements of modern AMI deployments result in increased system efficiency and reduced consumer cost.

From Meter to Smart Grid to Smart City

The AMI system is a complex network of numerous hardware and software components. However, AMI is a subsystem of a larger system, or a system-of-systems (SoS), that is the Smart Grid. This includes the power generation, transmission, and distribution systems which all have layers of complexity that include large mechanical hardware, sensors, monitoring software, and systems for advanced data analytics. Each of these system have system goals or objectives of their own, such as maximum output with optimal efficiency, high availability, reliability, etc. However, as we bring these systems together a larger purpose emerges which addresses the needs of a broader set of stakeholders, like the consumer community and local business economy. These goals also entail broader and potentially more critical objectives such as global sustainability, end-to-end reliability of service, resiliency, and national security.

The concept of systems-of-systems is captured by Krygiel on page 33 of [4]:

> … systems-of-systems are systems-of-interest whose system elements are themselves systems; typically these entail large-scale inter-disciplinary problems involving multiple, heterogeneous, distributed systems. These interoperating collections of component systems usually produce results unachievable by the individual systems alone.

As the market pushes for more end-to-end quality of service and accountability, the result is increased need for more interoperability between elements across several systems which scale to even larger and more complex systems-of-systems like the Smart City.

The concepts of cities or communities date back thousands of years when people started living together in groups and creating communities for various social benefits, improved life styles, security, and so forth. However, with the recent advances in technology, sensors, real-time data, and powerful analytical computation, city stakeholders are pressing for "smarter" city services. These utility services extend beyond providing power. Modern requirements include societal priorities such as clean water distribution, wastewater collection, transportation, healthcare, education, national security, social services, and such. To satisfy growing societal demands and consumer expectations, these services need to be provided with constantly improving features, such as reliability, resiliency, affordability, sustainability, and security.

The increasing number and complexity of system elements is magnified by the number and complexity of technologies. Additionally, the need for all these systems to be interoperable produces a great need for global standards to keep these elements compatible with each other and able to exchange information with each other securely and efficiently.

Role of Standards Within Smart Cities and Other Complex Systems

Technical standards in power distribution, measurement, and quality are important and have existed at the element level since the 1900s. Such technical standards are established norms or requirements that have been defined privately or unilaterally within companies or organizations. Standards can also be developed by organizations whose primary purpose is to develop, coordinate, promulgate, maintain, and interpret these standards. These organizations are typically referred to as Standards Developing Organizations (SDO) which typically have a diverse group of members from the industry which help develop these standards. In most cases, standards are voluntary, but can become mandatory if adopted by the government or business requirements.

The first metering standard was published in 1910 by the American National Standards Institute (ANSI), and was called the ANSI "C12 Code". This standard established acceptable performance criteria for the newly developed AC Watt hour meter, which was previously described. In 1931, the International Electrotechnical Commission (IEC) came out with Publication 43, "Alternating Current Watt-hour". The proliferation of standards (and standards organizations) is critical to the management and architecture of the modern "system of systems."

For example, thousands of electrical standards exist today, and have been developed by many SDOs. Within the IEC, there are hundreds of standards that are relevant to the AMI system. According to the *IEC Smart Grid Standards Map* [5], there are 70 standards that apply to the meter; 133 standards on the AMI Backhaul Network; 149 standards on the Network Interface Controller; 93 standards on the AMI Head End or UI. It's apparent that as elements advance in functionality and the number of elements of a system increase, so does the number of standards that are required to support the safety, interoperability and compatibility of the overall system-of-systems. This requires SDOs to shift from a component-driven mind-set to a "systems thinking" perspective. Additionally, SDOs must be structured organizationally to support increasingly cross-functional engagement.

SDO's are now addressing standards gaps for large, complex systems like the Smart City. These are systems that extend beyond just hardware and software components and into considerations of environmental and human behavior impact. Tackling such large systems can be an overwhelming endeavor for any committee. Nevertheless, these considerations must be addressed because developing standards in silos or with a component-only focus can result in a system that is broken and unable to fulfill its higher system level functionality. Modern systems-of-systems must understand the impact of their operations on neighboring elements, systems or users.

SDOs on Systems

Traditionally, SDOs have been structured organizationally into committees or groups that focus on specific components or technology. Over time, this structure has created silos between group engagements as well as in the standards that have been developed. Large and small SDOs like the International Organization for Standardization (ISO), the International Electrotechnical Commission (IEC), the International Telecommunication Union (ITU) and others have slowly evolved along with technology and systems. These organizations now have technical groups that are focused on broader systems such as Smart Grid, Smart Cities, Smart Manufacturing, and many others. The IEC in particular has established a "Systems Work" model that incorporates a continuous evaluation of emerging systems in the market, dedicated system committees, and a central system resource group.

As mentioned, the IEC "Systems Work" model contains System Evaluation Groups (SEG), System Committees (SyC), and a central Systems Resource Group, all of which are managed under the IEC Standardization Management Board (SMB). In the normal standards development process, as new systems start emerging in the market, IEC National Committees or other groups within the IEC can propose the formation of a SEG to the SMB.

The purpose of the SEG is to evaluate the system of interest and report to the SMB whether there is a legitimate need to form a Systems Committee. During the SEG evaluation, relevant system stakeholders are identified, and their needs analyzed. Thus, a scope of the system is developed along with preliminary reference architecture, and existing standards are identified that are in scope of the system. Based on the analysis, the SEG may propose to the SMB whether to move forward with a Systems Committee or abandon the effort.

For example, in December 2013 the SMB established SEG1 on Smart Cities. In case of SEG1 on Smart Cities, the report to the SMB proposed the formation of a Systems Committee on Smart Cities called the "Systems Committee on Electrotechnical Aspects of Smart Cities" which was approved in October 2015. The resulting Systems Committee (SyC) develops standardization interfaces and functional requirements by working across a matrix of relevant IEC Technical Committees (TC).

It's apparent that the SEG and SyC have an important responsibility of understanding the complexity and intricacy of the system. The concept of a Smart City is a systems-of-systems, and understanding such complexity requires a systematic approach. The IEC's response to this challenge is the Systems Resource Group (SRG). The SRG provides a systems approach or guideline in evaluating such complex systems. The SRG also provide tools, techniques, and other services to support the SEGs and SyCs. The IEC is one of the very first SDOs to form such a committee.

System Committees and System Evaluation Groups are comprised of subject matter experts with a diverse range of backgrounds and work experience. Unfortunately, only a few of the members have direct experience in systems thinking and analysis. Without a systematic approach, groups can find themselves over-whelmed and lost in scope due to the magnitude and complexity of such systems-of-systems. Therefore, it is crucial to have a standard systematic approach with tools and techniques that are consistent across the organization to thoroughly, effectively, and efficiently examine and understand the complex of the system in focus. An outcome of these exercises enables SDO committees to better identify gaps in the system where standards are needed to bridge elements of the system and create interoperability.

Systems Approach

There are many system approaches and guidelines available today. However, one of the popular and widely used guidelines comes from the International Council on Systems Engineering (INCOSE) [6] which is consistent with ISO/IEC 15288 [7]

> man-made, created and utilized to provide products and/or services in defined environments for the benefit of users and other stakeholders. These systems may be configured with one or more of the following system elements: hardware, software, data, humans, processes (e.g., process for providing services to others), procedures (e.g., operator instructions), facilities, materials and naturally occurring entities. In practice, they are thought of as products or services.
>
> The perception and definition of a particular system, its architecture and its system elements depend on an observer's interests and responsibilities. One person's system-of-interest can be viewed as a system element in another person's system-of-interest. Furthermore, a system-of-interest can be viewed as being part of the environment of operation for another person's system-of-interest.

With this perspective, the Smart City is a "man-made" system that serves to benefit a wide range of stakeholders including residential citizens, commercial and public service providers, government officials, and many others who have an interest and responsibility in the city. However, as described previously, the Smart City is beyond just "a" system but a systems-of-systems and has life cycles [7]:

> Life cycles vary according to the nature, purpose, use and prevailing circumstances of the system. Each stage has a distinct purpose and contribution to the whole life cycle and is conserved when planning and executing the system life cycle. … The stages thus provide organizations with a framework within which organization management has high-level visibility and control of project and Technical Processes.

However, in "Smart Cities" there is no true "organization management" but a collaborative quilting of various stakeholders that are driving the direction of the system. In this context, the citizens of the city play a major role. The generic "life cycles" as defined by the standard includes those described in Table 3.1 [6].

Table 3.1 INCOSE life-cycle stages

Life-cycle stages	Purpose
Exploratory research	Identify stakeholders' needs; explore ideas and technologies
Concept	Refine stakeholder's needs; explore feasible concepts; propose viable solutions
Development	Refine system requirements; create solution description; build system; verify and validate system
Production	Produce systems; inspect and verify
Utilization	Operate system to satisfy users' needs
Support	Provide sustained system capability
Retirement	Store, archive, or dispose of the systems

ISO/IEC 15288:2008 [7] states

> Every system has a life cycle. … A system progresses through its life cycle as the result of actions, performed and managed by people in organizations, using processes for execution of these actions.
>
> The life cycle model comprises one or more stage models, as needed. It is assembled as a sequence of stages that may overlap and/or iterate, as appropriate for the system-of-interest's scope, magnitude, complexity, changing needs and opportunities.

In the case of a Smart City that is being analyzed by a SDO, some of the stages do not apply. A SDO committee is not necessarily "engineering" a Smart City, rather it is evaluating an existing operating system with the objective of identifying gaps or areas of concern in the system where standards can help fulfill or resolve. In order to accomplish this goal, SDOs need to have a holistic view of the system. This includes understanding who the stakeholders are, what their needs are, the elements within each system, and existing standards that impact the system and its elements. Within a Smart City, the various systems are constantly evolving and changing as new technologies are introduced. Additionally, needs and requirements are continuously changing. This makes the process quite challenging, but with the use of tools and methodologies it can be manageable.

Collaboration, Traceability, and Iteration

There are three essential concepts to understand and apply that are crucial to successfully developing standards for complex systems like Smart Cities. These concepts are: collaboration, traceability, and iteration.

Every city, region, and country has different needs, drivers, and regulations when it comes to Smart Cities. This creates a unique challenge for SDOs who have to develop standards that are applicable across the board which is why collaboration is very important. Every representative needs to be heard in order to gather a comprehensive landscape of the needs from each region. Of course, one standard will not fit every need of every city, but a due diligence effort has to be made to understand the commonalities and create standards that play a key role in the building blocks of a Smart City initiative.

There are some basic collaboration activities that should be practiced to ensure everyone's voice is heard especially in a committee that is very diverse. There are individuals who are extroverts and more assertive, therefore their voices will always be heard loud and clear. However, for those who are on the other end of the spectrum or come from cultures that are less assertive require different means of gathering inputs outside of just roundtable discussions. This requires creating a safe environment where everyone is free to provide their inputs without any negative push-back but rather create constructive discussions. Committees can start by making simple efforts like ensuring everyone is providing inputs. Other activities like using "written responses" on whiteboards or feedback cards are also good for gathering inputs.

The second important concept to practice is traceability. This means ensuring that all activities, objectives, and deliverables trace back to the committee's identified goals/needs. For example, a Smart City has many broad initiatives like sustainability, accessibility, livability, etc. However, when decomposed into its smaller parts there are a larger number of objectives to meet these initiatives like "sustainable power supply to the healthcare providers" or "accessible transportation for elderly citizens." OnceSDO committees have identified key objectives it becomes an art and discipline to ensure all the subsequent activities like analyzing stakeholders, developing use cases, creating architectures and such are traceable back to the committee's original objectives to ensure focus. With large complex systems, scope creep and analysis paralysis becomes very common and can drive the group into a spiral and off the planned roadmap. Therefore, each analysis activity and its outputs are traceable back to the defined objective.

Finally, iteration is the understanding that as working committee goes through the stages of the system life-cycle process that it may require more than one iteration of the cycle or possibly going back a stage when new findings is discovered in the current stage. Often, committees get caught in the activities in one stage and drill down too deep that they lose track of going to the next stage. Its best practice to go through the first iteration of the life cycle at a high-level to help build a general understanding, then in the second iteration go down to the next level of detail. This helps prevent analysis paralysis and also helps the group generate a rapid foundational perspective and understanding early. As mentioned, iteration also means having to go back a stage when new discoveries are made that may have been missed in the previous stage. For example, sometime after doing a Use Case Analysis the group may identify several stakeholders that were not identified in the previous Stakeholder Analysis stage. Therefore, the group should go back and access whether these new stakeholders should be added and understand how this impacts any other aspects of the objective.

Concluding Remarks

In summary, a Smart City is a very complex system comprised of components that have evolved with advanced features and capabilities that are interacting with so many other elements, creating systems within systems. Developing standards for such large systems require a systematic approach especially when there are existing standards at the component level that have to evolve in order to carry out more system level objectives. Applying a systems approach provides a guideline and a systematic methodology for consistency and holistic understanding of the entire system-of-systems. The systems approach should also be applied with a spirit of collaboration to embrace diversity, as all cities and countries are different. Iteration and traceability are also important to guide the process and ensure thoroughness, effectiveness, and maintaining focus. Finally, developing standards for large complex systems like a Smart City can seem like a daunting task, and can seem as if you need to boil the ocean. However, with a defined systems approach there is light at the end of the tunnel.

References

1. Tesla N. http://www.history.com/topics/inventions/nikola-tesla. Accessed Mar 2017
2. The History of Electricity Meters June 28, 2006. https://www.metering.com/features/the-history-of-the-electricity-meter. Accessed Nov 2016
3. Advance Metering Count by Technology Type 2007 Through 2015, U.S. Energy Information Administration. https://www.eia.gov/electricity/annual/html/epa_10_10.html. Accessed March 2017
4. Krygiel AJ (1999) Behind the wizard's curtain. CCRP Publication Series, Washington
5. Smart Grid Standards Map, International Electrotechnical Commission. http://www.smartgridstandardsmap.com. Accessed Nov 2016
6. Systems engineering handbook: a guide or systems life cycle processes and activities. International Council on Systems Engineering (INCOSE), 2011
7. Systems and software engineering—System life cycle processes. ISO/IEC 15288:2008. ISO, Geneva

Part II
Technology & Architecture

Chapter 4
The Smart Grid: Anchor of the Smart City

George Koutitas

Introduction

The power grid network is the backbone of the city. It delivers the energy required for everyday operations. Since smart cities are built on the two main pillars of sustainability and clean energy, the smart grid can be considered to be a fundamental element of the Smart City. The smart grid integrates power grid technologies with Information and Communication Technology (ICT) systems to create an energy distribution system that:

- Enables the faster penetration of renewable energy sources (RES) by managing the consumption according to time-variable RES production.
- Can self-heal in extreme conditions, minimizing the number of outages and maximizing the Quality of Services (QoS).
- Creates a "human-centric-grid" where customer-centric applications and new business models in the competitive economy places the end customer as the epicenter.

The smart grid is a *system of systems* and a *business of businesses*. The system of systems involves the orchestration of power grid components such as (a) energy production units (solar, thermal, etc.), and (b) energy distribution units (transformers, cables, etc.) as well as (c) ICT devices for data transfer (networks, modems, etc.) and command flow (smart meters, load controllers, etc.). The business of businesses involves the collaboration of a great diversity of players such as utilities, energy producers, product innovators, IT vendors, and application providers. The

G. Koutitas (✉)
Gridmates Inc., Austin, TX, USA
e-mail: george@gridmates.com

G. Koutitas
Texas State University, San Marcos, TX, USA

S. McClellan et al. (eds.), *Smart Cities*, DOI 10.1007/978-3-319-59381-4_4

following sections of this chapter present the elements of the smart grid, starting with the drawbacks of the conventional power grid architecture which led to modernization and produced the smart grid. The chapter ends with the presentation of the most modern applications and future trends that help smart cities enhance their services and create a network of engaged citizens.

Existing Architecture

Players, Regions, and Markets

The power grid is the largest network in the world and it connects producers and consumers of energy. The current network architecture in the United States is based on a centralized system and is segmented in different regions. The North American Electric Reliability Council (NERC, now called the North American Electric Reliability Corporation), established in 1968, ensures a reliable and secure network operation across the country by orchestrating local grids that were created in various regions of the U.S. The U.S. power grid is segmented in 8 regional organizations that provide services to different geographical regions such as the Florida Reliability Coordinating Council (FRCC), Midwest Reliability Organization (MRO), Northeast Power Coordinating Council (NPCC), Reliability First Corporation (RFC), SERC Reliability Corporation (SERC), Southwest Power Pool, RE (SPP), Texas Reliability Entity (TRE), and Western Electricity Coordinating Council (WECC) [1]. The regional organizations are referred to as Independent System Operators (ISOs) and Regional Transmission Organizations (RTOs) and are part of a national standard advocated by the Federal Energy Regulatory Commission (FERC).

Electric services are provided to approximately 134 million customers by 3000 electric distribution utilities using more than 200,000 miles of transmission and distribution lines and leveraging the power production of more than 7200 power plants and generating facilities in the U.S. [2, 3]. The electric distribution utilities are categorized as Municipality Owned Utilities (MOUs), Investor Owned Utilities (IOUs), electric Cooperatives (COOPs) and Retail Energy Providers (REPs). MOUs and COOPs are public companies that may incorporate production and distribution assets and they provide electric services to urban and rural environments, respectively. IOUs are private companies that may incorporate production facilities and are usually found in densely populated areas. Finally, REPs are private companies that connect consumers to producers by buying energy from the wholesale market and reselling to the end users. REPs operate in deregulated market areas.

Almost all electric utilities participate in the energy market where selling and buying of electricity takes place. Utilities buy electricity when their own production is not adequate to satisfy the demand of their service territory. The energy market is comprised of *wholesale* and *retail* segments and is divided into *regulated* and *deregulated* markets [4]. In regulated markets, electric utilities operate in a vertical manner and own the generation, distribution and retail sectors of the business. In this market, infrastructure is the asset. This type of market limits the participation of other business players and minimizes competition. In deregulated markets, the electric utility competes with other business players and needs to attract its own customers by offering innovative services and competitive electric prices. In this market, customer engagement is the asset. The commodities transacted in the wholesale market are energy, power, ancillary services and congestions or losses, and these define the final price. Energy (in MWh) represents the actual commodity that is consumed by the end customers. Energy markets trade net-generation in time intervals such as 5, 15, and 60 min. The main players in the wholesale market are the *Independent System Operator* (ISO), *Market Operator* (MO), *Transmission System Operators* (TSO), *Generators*, *Retailers*, and *Consumers*. The ISO manages the spot market in different regions and provides an operational umbrella for the market. The MO clears and settles electricity transactions. The TSO is a controlling agency which coordinates supply and demand (e.g., the dispatch of the generating energy units to the demand). The generators produce electricity and bid it to the market. The retailers purchase large amounts of electricity and resell it to consumers via supply contracts.

Network Architecture

The existing power grid architecture is centralized in nature and it delivers the energy produced from large generation stations to the end consumer. The four elements found in the delivery of electric services are: (1) Generation station or power plant, (2) Transmission network, (3) Distribution network, and (4) Consumer (end user). These elements are presented in Fig. 4.1.

Generation Stations Power plants convert different forms of energy to direct current (DC) electricity. Natural resources include fossil fuels, nuclear power, or

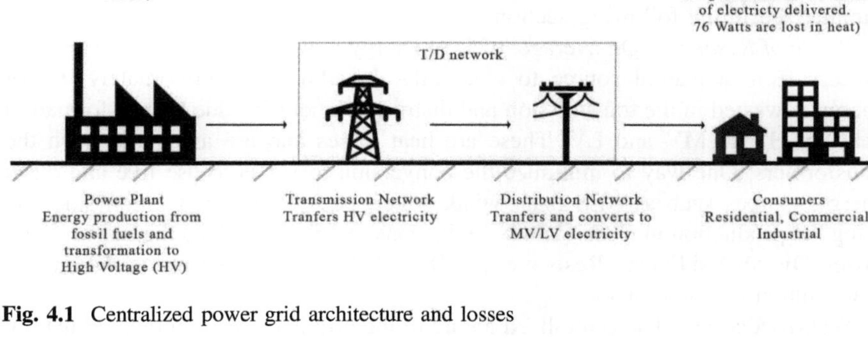

Fig. 4.1 Centralized power grid architecture and losses

renewable energy sources (RES). In recent years, RES have attracted a lot of momentum since they minimize the approximately 62% conversion losses [3] observed in fossil fuels. The electricity produced via this transformation is then supplied to the transmission network for transportation to the distribution network, and finally the consumers. Power plants are usually deployed near the location of the natural energy source, for example, the coal mining regions, to minimize the costs of transportation of the natural resource to the generation unit. The main drawback is that large transmission networks are necessary to transfer the produced electricity to the end consumer. This is one of the challenges addressed by the smart grid, which brings the production and consumption in close vicinity.

Transmission Network The transmission network processes the voltage generated from the power plant using a "step up" conversion to high voltage (HV) using transformers. This process is necessary for transporting electricity over large distances. HV transmission is preferable because it involves lower current on the transmission cables, which results in less heat through resistive loss. Three-phase alternate current (AC) transmission is used because HV AC signals can be easily manipulated to optimize transmission and distribution networks. Depending on the distance and location, HV values used for transmission may be 756, 500, 345, 230, or 138 kV.

Distribution Network The distribution network is the last part of the network and delivers electricity to the consumer. The distribution network transforms the HV electricity of the transmission network to medium voltage (MV) for industrial/commercial consumers (4, 13, 26 or 69 kV) or low voltage (LV) for residential consumers (120 or 240 V). The conversion from HV to MV or LV is important since it adapts the distributed power to the standard or conventional power values expected by the consumer and the appliances.

Drawbacks of the Existing Power Grid

The existing centralized power grid architecture increases the losses and inefficiencies and creates some critical business concerns. The main drawbacks are summarized in the following sections.

Waste of Resources On average, 62% of energy is wasted during the conversion process from a natural source to electricity. In addition, approximately 9% of energy is wasted in the transmission and distribution network due to transformation between HV to MV and LV. These are heat losses and are associated with the transformers. One way to minimize the conversion losses is to use free and clean energy sources such as RES (sun, wind, water, etc.) for the generation side and bring the production in close vicinity to the consumption to reduce the transmission losses. Distributed Energy Resources (DER) and Distributed Generation (DG) is the key solution to this problem.

Security/Outages The centralized nature of the grid is more vulnerable to natural disasters, terrorist attacks, or even cyberattacks. To strengthen the operation of the power grid, it is important to migrate from a centralized architecture to a distributed grid network that can self-heal in response to external perturbations and attacks. This requires real-time monitoring and control capabilities achieved by deploying intelligence at the edge of the network.

Adaptation to RES Renewable energy sources (RES) produce a time-variable generation of electricity that depends on environmental conditions (e.g., wind speed, solar intensity, etc.). One way to alleviate the potential problems of outages is to manage the consumption according to the time-variable capacity offered by RES. This is performed through Demand Side Management (DSM) and Demand Response (DR) programs, which are available via smart grid technologies.

Passive Nature of Consumers One of the most vital elements of a dynamic system is the active participation of its users. The existing centralized grid architecture supports one-way communication flow between the electric provider and the end consumer. Consumers hold a passive role without being able to participate in the grid operation. The smart grid enables a bidirectional data and command flow between the consumer and the electric provider. Consumers can interact with their connected devices, learn about their energy consumption or even provide capacity to the power grid network in the form of solar energy or "negawatt" energy from energy savings. The smart grid strengthens the role of the consumer and transforms passive consumers to active prosumers (coming from the combination of the words producer and consumer of energy). This is very important for the Smart City environment since it is also based on a new active citizen model rather than the current passive citizen model.

Business Models In many cases, the vertical operation of utilities limits the competition and does not allow third parties to enter the market. This has as a consequence the lack of innovation and lack of the development of customer-centric services and Business to Consumer (B2C) products. The smart grid creates a fertile foundation to allow new business models to blossom.

Evolution Toward a Smarter Grid

The power grid is experiencing a transformation similar to that of the mobile telecommunications industry. It is being transformed from a utility-centric network, to a *customer-centric network*, from a centralized network with passive customers to a *distributed system* of smart devices with *active prosumers* who can manage certain aspects of the operation of the grid [6]. Smart metering technologies and new applications create intelligence at the edge of the network, following the mobile industry's philosophy with smart phones [7]. In addition, new business models and players have emerged focusing on application providers and product innovators. Finally, electric utilities transition from *asset owners* to *enablers of services* where the asset is the data and *not* the infrastructure. This follows the same principles as in smart cities, where citizen engagement and service offerings are the assets. The smart operation of the power grid includes a sequence of events that take place in a fraction of a second, including:

- Data is collected from the edge of the network (smart meters, appliances, etc.),
- Data is converted to information at the core network (databases, third party application companies, utilities, vendors),
- Information is processed and converted to knowledge and intelligence, and
- Intelligence is converted to command flow to enable automation and interaction at the edge of the network,
- Energy is delivered through a large number of Distributed Energy Resources (DER).

These processes require the orchestration of new business partners and technologies that are discussed in the following sections.

Utility of the Future

The traditional business models of the utility industry are challenged by new concepts that transform utilities to product innovators, network managers, and enablers of services [5]. The key reasons for this transformation are the *new clean energy technologies* such as RES and distributed storage, the *innovative service providers* that can be regarded as nontraditional competitors, the *modern customer behavior* that is based on the world of mobile and infinite connectivity, the new *regulatory pressures*, and of course the *rising costs of electricity*. A similar transformation was also encountered in the telecom industry (1995–2005 +), where the traditional infrastructure-oriented telephone network business was challenged by the data-driven content and multimedia business.

The following significant changes are expected to occur by the year 2030 [8]:

- 15% of demand reduction will be feasible through smart thermostats
- more than 57% of residential utility customers are expected to own connected home devices and smart appliances
- $40 billion in revenue reduction is expected from load reductions.

It is clear that traditional energy providers should transition from a conventional business mentality such as "obligation to serve" to a new mentality such as "commitment to optimize." The utility of the future should focus on low-carbon and clean energy production, become optimizers of various distribution platforms, and enable new customer-centric services and product innovation. New business models such as the *Demand Response* business model, *Energy Management System* and storage business model, *Community Solar* business model, and *Business to Consumer* model emerge to enhance the existing ecosystem. The utility of the future is transitioning to a partner-of-partners model and product innovators aiming to leverage the roles of the residential and commercial customers.

Utility Customer Beyond 2020

One of the main forces driving the transition of the utility business is the modern customer behavior and future roles. The utility customer beyond 2020 is expected to have some critical changes in terms of the customer behavior and customer expectation [9]. The future utility customer is *digital*, *connected*, and *social*:

- *Digital* Approximately 95% of utility customers are expected to be tech savvy and interact with their utility using their smart phones. For utilities, this means that they should invest in mobile bill payment technologies and cloud-based solutions
- *Connected* It is expected that every home will have 50 connected devices. From smart phones to smart thermostats and health information systems. Utilities should invest in product innovation and integrations among various distribution platforms
- *Social* Millennials and Generation X customers have become the largest buying sector in the utility space. They are all connected via social media and have strong environmental and social consciousness. Utilities should implement smart social responsibility initiatives and leverage the connectivity of the customers using their social media.

Smart Grid Elements

The smart grid is based upon significant technological achievements in the Information and Communication Technologies (ICT) and energy sectors. The backbone of innovation is the integration of ICT with the energy technologies that create new services and business models. This integration enables the real-time command and data flow between the nodes of the power grid network. This real-time interaction enables the development of new concepts and elements in the smart grid architecture. The most significant elements are the Smart Metering layer, Prosumers, Microgrids, and Virtual Power Plants (VPP) as presented in Fig. 4.2.

Smart Metering Layer The smart metering layer, part of the overall ICT infrastructure of the smart grid, provides the fundamental intelligence. It is responsible for real-time command and data flow between the different network nodes. The most critical nodes are the smart meters and load controllers that measure and transfer energy consumption/production data to the utility and enable load management through on/off command flow. The smart metering network is also called Advanced Metering Infrastructure (AMI), and its deployment in the utility service territory is considered as the first step toward network modernization (smart grid 1.0). In most cases, IT vendors help utilities deploy these networks in a Business to Business (B2B) type of collaboration since utilities do not have the

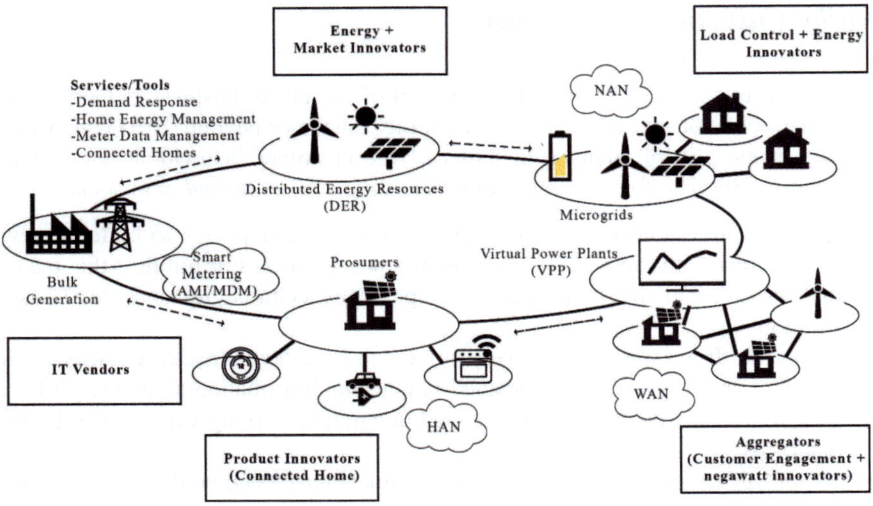

Fig. 4.2 The smart grid architecture, elements, and business entities

relevant IT expertise. AMI captures energy consumption/production data from the consumer side and transfers the data to central databases. In these databases, Meter Data Management (MDM) systems process raw data, converting it into information and thus intelligence. MDM and AMI are usually coupled to the Customer Information System (CIS) that provide billing, but they are also connected to third party application providers that leverage big data and provide more sophisticated customer-centric applications in a Business to Consumer (B2C) type of collaboration [10]. Data transmission utilizes wireless and wired communication networks. The main challenge of the deployment of such communication networks is the provision of coverage from large geographical areas (High Voltage or Medium Voltage) to indoor environments (Low Voltage part of the grid). Within homes, there are Home Energy Management systems (HEMs) that manage and connect smart appliances and smart meters to a home Access Point (AP) and software Agent through Home Area Networks (HAN). HANs usually implement mesh network protocols such as Zigbee, IEEE 802.15.4 and 6LowPAN [11]. HEMs are connected to the utility through Wide Area Networks (WAN) or Neighborhood Area Networks (NAN). The most commonly used protocols are the Internet Protocol (IP) and Transport Control Protocol (TCP) which constitute the backbone of the Internet of Things (IoT) and Machine to Machine (M2 M) communications. Very recently, low latency M2 M communications inspired the creation of the concept called "tactile Internet" [12] that enables secure, reliable, high availability with low latency (1 ms response) Internet connections among smart devices. The real-time data and command flow between smart meters and controllers enable new services such as Demand Response (DR). DR provides load management according to time-variable production or pricing signals or peak power conditions. DR is usually implemented using thermostatic loads such as air conditioning units.

Prosumer: The role of electricity consumers has dramatically changed in recent years, from passive to active consumers. The term active is related to the fact that there is a bidirectional communication and interaction between the residential consumer and the electric utility. The active nature of the consumer helps the everyday operation of the grid in terms of responsiveness during critical hours. Active consumers have further transformed to prosumers, which is a combination of the words "consumer" and "producer." A prosumer may produce clean energy from renewable energy sources (RES) such as rooftop solar panels or solar tiles. The prosumer may also produce "negawatt" energy (or capacity) from energy management using smart thermostats and Demand Response (DR) techniques. This negawatt capacity is a virtual production since it reduces the requirements for energy generation. Finally, the prosumer can store energy using home energy storage battery banks.

The prosumer has two states of operation according to the relationship between the production $p(t)$ in Watts and demand, $c(t)$ in Watts. During the production state, the available capacity is larger than the demand, $p(t) > c(t)$ and the prosumer provides energy to the grid. It should be noted that the production can be the sum of the solar and negawatt capacity indicated as $p(t) = s(t) + n(t)$. For example, when the prosumer has a solar panel and during the mid-day the production is larger than the overall consumption of the residence, then the prosumer is in production state. During the consumption state, the consumption is larger than the demand $c(t) > p(t)$ and grid-imported energy is necessary. Due to costs associated with energy storage and solar panels, the majority of prosumers during a billing period require grid-imported energy to satisfy their net demand. However, this condition is changing and off-grid residential customers have emerged since battery storage is now available in competitive prices. For example, a 1-bedroom home that has a daily consumption of 10 kW requires solar panel of 5 kW and a battery capacity of 14 kW to have 100% energy independence with overall cost below $20,000.

Microgrids: Microgrids bring consumption and production in close vicinity. A microgrid is a small scale version of the grid, but the production units and the loads can be within the same medium voltage (MV) or even low voltage (LV) network where sometimes direct current (DC) transmission can be achieved. A microgrid is a group of interconnected Distributed Energy Resources (DER), usually based on RES and large storage facilities (battery banks) and smart loads, such as smart homes. The smart homes are prosumers since they can offer "negawatt" capacity through energy management techniques. The key benefit of a microgrid is that it minimizes energy conversion losses since the production is based on RES. It also minimizes the transport losses since consumption and production can coexist within the same MV or LV network without the need for voltage transformation. The microgrid is controlled by a central software agent that monitors consumption and production, and makes decisions according to real-time characteristics. These decisions concern the import of energy from the grid or the broadcast of on/off commands to the loads of the microgrid for load reduction (for example reduction of smart thermostat temperatures). The microgrid can operate in an "island" mode if it is totally disconnected from the grid. In this case, the total consumption should always be smaller than the production. Examples of microgrids are found in [13, 14]. Microgrids are important elements of smart cities since they can enable various neighborhoods or groups of buildings to operate using their own clean energy infrastructure.

Virtual Power Plants (VPP) VPPs are the most recent advancement of the smart grid. It is hard to describe a VPP, but one can consider that a VPP is an amorphous coalition of DER and prosumers (DG), controlled by a single entity so as to give a reliable overall power supply and participate to the energy market [15, 16]. An important characteristic of a VPP is that its Distributed Generation (DG) units are not required to be connected to the same MV or LV network since the production of the VPP is not used for local consumption. The VPP operates as an aggregator to enable market participation for their members (prosumers and DGs). VPPs are the foundation upon which the model of *Transactive Energy* is built. More details about Transactive Energy are presented in the last section of this chapter.

Vehicle to Grid (V2G): Electric Vehicles (EVs) have presented a tremendous market penetration during the recent years due to the reduction of battery cost and the deployment of a dense network of EV chargers. In 2005, the battery storage cost was approximately $1300/kW, whereas in 2015 it reduced to $400/kW [17]. Fully electric EVs have a mileage range that varies from 60 to 300 miles and are powered by a battery of 20–100 kW. If one considers that a large number of EVs with large storage capabilities are in a Smart City environment, it is obvious that this Distributed Storage (DS) can provide important benefits to grid operation. Similar to the solar roofs and photovoltaic systems, the Vehicle to Grid (V2G) supports a two-way power flow to and from the grid. An EV owner may consume energy from the grid to charge the EV, or may provide energy to the grid. From the grid perspective, a large number of EVs may be regarded as a distributed battery bank or a Distributed Generation (DG) system.

Standards

A great challenge of the implementation of smart grid vision is the interoperability of the different types of technologies used. The smart grid is the integration of new and existing ICT and energy systems and interoperability is of vital importance. Initiatives such as the smart grid Interoperability [18, 19] provide the necessary roadmap for system integration. The smart grid is a very complex and distributed system, and each entity utilizes different data structures, communication networks, and information systems which need to be interconnected and orchestrated.

Let us consider the following case. A smart meter of a residential home is part of the AMI and needs to be connected to the MDM. The connection is performed through a wireless or wired network. The MDM should also be able to exchange data with the CIS for billing purposes. Unfortunately, when various vendors utilize different systems, they also use different data formats and software bridges. This results in a collection of interoperability issues. Interoperability is achieved with the use of appropriate protocols and standards. There are hundreds of protocols and standards divided among the various domains of the system such as the customer, the service provider, the generation, the transmission, the distribution, and the operations domains. The most widely used interoperability protocols are listed in Tables 4.1 and 4.2.

Table 4.1 Some of the most commonly used protocols and standards for smart grid interoperability, facing the utility [6, 18]

Standard/protocol	Description
ANSI C12.18	Revenue metering for end devices
C12.20	Transport of measurement device data over telephone networks
NSI/CEA 709 and Consumer Electronics Association 852.1 LON protocol suite	This is a general purpose local area networking protocol in use for various applications including electric meters, street lighting, home automation, and building automation
IEEE 1815 (DNP3)	This standard is used for substation and feeder device automation, as well as for communications between control centers and substations
IEC 61850 suite: communication networks and systems in substations	This standard defines communications within transmission and distribution substations for automation and protection
IEC 61968/61970 suites	These standards define information exchanged among control center systems using common information models. They define application-level energy management system interfaces and messaging for distribution grid management in the utility space
MultiSpeak	A specification for application software integration within the utility operations domain; a candidate for use in an Enterprise Service Bus. It is widely used for integration with MDM and CIS
Open geospatial consortium geography markup language (GML)	A standard for exchange of location-based information addressing geographic data requirements for many smart grid applications
EMIX (Energy Market Information eXchange)	EMIX provides an information model to enable the exchange of energy price, characteristics, time, and related information for wholesale energy markets, including market makers, market participants, quote streams, premises automation, and devices

(continued)

Table 4.1 (continued)

Standard/protocol	Description
Internet protocol suite, request for comments (RFC) 6272, Internet protocols for the smart grid	Internet Protocols for IP-based smart grid networks IPv4/IPv6 are the foundation protocol for delivery of packets in the Internet network. Internet Protocol version 6 (IPv6) is a new version of the Internet Protocol that provides enhancements to Internet Protocol version 4 (IPv4) and allows a larger address space
SAE J2836/1	This document establishes use cases for communication between plug-in electric vehicles and the electric power grid, for energy transfer and other applications: use cases for communication between plug-in vehicles and the utility grid
Security profile for advanced metering infrastructure, v 1.0	Advanced Security Acceleration Project—Smart Grid, December 10, 2009. This document provides guidance and security controls to organizations developing or implementing AMI solutions. This includes the meter data management system (MDMS) up to and including the HAN interface of the smart meter
IEC 62351: Power systems management and associated information exchange—data and communications security	Open standard that defines security requirements for power system management and information exchange, including communications network and system security issues, transmission control protocol (TCP) and manufacturing messaging specification (MMS) profiles, and security for inter-control center protocol (ICCP) and substation automation and protection. It is for use in conjunction with related IEC standards, but has not been widely adopted yet
IEEE 1686-2007	The IEEE 1686-2007 is a standard that defines functions and features to be provided in substation intelligent electronic devices (IEDs) for critical infrastructure protection programs. The standard covers IED security capabilities including the access, operation, configuration, firmware revision, and data retrieval
SOAP (Simple Object Access Protocol)	It is used for exchanging structured information in the implementation of web services in computer networks. It uses XML for its message format, and relies on application layer protocols (HTTP) or (SMTP)
XML (Extensible Markup Language)	Is a markup language that defines a set of rules for encoding documents in a format that is both human-readable and machine-readable

Table 4.2 Some of the most commonly used protocols and standards for smart grid interoperability, facing the customer [6, 18]

Standard/protocol	Description
Open automated demand 2.0 response (OpenADR)	The specification defines messages exchanged between the demand response (DR) service providers (e.g., utilities, independent system operators (ISOs) and customers for price-responsive and reliability-based DR
Smart energy profile 2.0	Home Area Network (HAN) device communications and information model
OpenHAN	A specification for home area network (HAN) to connect to the utility advanced metering system including device communication, measurement, and control
IEEE 802.11 (WiFi)	IEEE Standard for information technology–telecommunications and information exchange between systems local and metropolitan area networks
IEEE 802.15.4 (Zigbee)	IEEE Standard for local and metropolitan area networks
IEEE 802.16	IEEE standard for air interface for broadband wireless access systems
3G/4G/LTE	Wireless cellular networks for data transfer
THREAD	Thread uses 6LoWPAN, which in turn uses the IEEE 802.15.4 wireless protocol with mesh communication, as does ZigBee and other systems. Thread however is IP-addressable, with cloud access and AES encryption. It is widely used for smarthome automation systems

Transition to an Application Development Platform

The orchestration of diverse systems that provide real-time monitoring and management create the foundations for the "Internet of energy." The smart grid follows an evolution similar to that of the mobile telecommunications industry, where the smart grid converges toward an application development platform. The first generation of the smart grid, smart grid 1.0 was responsible to deploy ICT and smart metering networks and enable real-time monitoring and management. The second generation, smart grid 2.0, deployed DER and clean energy technologies and enabled new entities such as the prosumers, microgrids and VPPs to evolve. We are now experiencing smart grid 3.0 where customer-centric applications and connected home devices play a critical role.

The smart grid transforms to an application development platform where application providers offer in a Business to Business (B2B) or Business to Consumer (B2C) manner applications for energy management and customer engagement. The real asset is not the infrastructure but the data ownership and customer engagement. To enable such a transition, utilities in the smart grid 3.0 focus on cloud-based solutions that leverage open data and Application Programming Interfaces (APIs).

Open Data

Open data and APIs are key enablers of the customer-centric smart grid, and smart cities follow this paradigm. A great variety of Smart City APIs and open data is presented in [20]. A great diversity of datasets is now easily accessible by third party application developers or even smart citizens who can develop and offer to the public their own apps. For example, the City of Austin has made available real-time data for traffic cameras, CO_2 emissions, and green buildings statistics. Anyone can access the data and create applications and services for the benefit of the community.

In a similar manner, open data and APIs related to the smart grid make possible the efficient development of new customer-centric services. The most widely used smart grid open data and APIs is summarized below.

Energy Information Administration (EIA): This initiative of the U.S. Department of Energy disseminates information related to energy production, consumption, and management and informs the public about the interaction of the energy with the economy and the environment [21]. A great diversity of databases is hosted by the EIA providing access to sales of electricity, revenues and prices, retail cost of electricity, demand and emissions, energy usage analysis, RES production and projections of energy-related values for different geographical regions.

Green Button The Green Button [22] is an open database and API offered by the U.S. Department of Energy to tackle the problem of integration with utility and smart meter databases. Green Button works as an independent third party database hosting residential smart meter data. An application provider can go directly to the Green Button database and obtain access to residential smart meter data. This access accelerates the evolution of new applications related to big data and energy analytics. Green Button allows data to be exchanged between three entities: Retail Customer, Data Custodian, and Third Party. The Retail Customer is any customer (residential, commercial, industrial) served by a utility which offers smart meter data to the Green Button database. The Data Custodian is the enterprise (utility or energy provider) holding the smart metered data. Finally, the Third Party is an application provider willing to access smart meter data to develop and offer new services to the retail customer. Green Button offers two types of APIs for data sharing: the Download My Data (DMD) API and the Connect My Data (CMD) API. The DMD service connects the retail customer with the data custodian. The CMD API provides application developers with an automated technique to access consumer energy information while maintaining security and privacy.

Orange Button is the Orange Button is an initiative of the U.S. Department of Energy that provides access to solar data [23]. Orange Button follows the principles and success of the Green Button APIs but focuses on the solar industry. Despite the fact that currently there is no open API to support solar data exchange, Orange Button will provide two important data platforms for solar data sharing: The Solar Data Translation Platform (SDTP) and the Solar Data Exchange Platform (SDEP). The SDTP is focused on translating data structures into standardized formats while the SDEP is focused on connecting industry to the standardized data, improving access, and advancing the solar marketplace. The eventual goal for the Orange Button data platform is to enable a marketplace for solar data exchanges, purchases, and connections.

PVWatts: *API* This is an open API and initiative of the National Renewable Energy Laboratory (NREL) that enables third party application developers to obtain access to environmental and solar conditions necessary for the computations of solar energy production [24]. The service estimates the performance of hypothetical residential and small commercial PV installations based on actual weather data. The estimates are based on a simulation model, called the PVWatts calculator, which processes input parameters related to the PV installation and weather data to provide an output describing the theoretical performance and energy production of the system in kW.

Connected Home: *APIs* A large number of private companies who are product innovators offer APIs that allow third parties to obtain access to their products. This helps the companies accelerate the adoption of their products from end users. Characteristic examples are microinverter APIs [25] that provide measured solar energy production values and Smart Thermostat APIs [26–28] that allow access to data as well as control.

Cloud-Based Services

Access to Open Databases and Application Programming Interfaces (APIs) is usually performed with the use of web technologies and cloud-based solutions. Since most of the systems from different vendors and utilities are connected to the cloud, web services and APIs provide the required tools for the interconnection of the systems. For example, an application provider may require access to the Meter Data Management (MDM) of the back office of the utility. The application provider may collect data from the MDM using customized "software bridges," web services, or APIs, and then process the data to deliver useful insights to the end user via web-based or smartphone applications.

A web service is a collection of open protocols and standards used for the exchange of data between various network nodes. For example, a web service may be used to connect a smart thermostat or the Home Energy Management System (HEMS) of a residential customer to the MDM system of the utility. The web service uses IP-based, open communication standards for the connection of the network nodes [29]. An Application Programming Interface (API) is a set of routines, data structures, and protocols which support the building of applications that require the orchestration of different systems. An API provides a client (a software system) with the ability to execute specific actions or request specific data from a server (a different software system). The client and server may be different computer nodes in different locations, or they may be different pieces of software on the same physical computer node. The most commonly used actions in a web-based

API are the commands GET/POST/PATCH/DELETE which are basic actions supported by Hyper Text Transfer Protocol (HTTP). For example, a mobile app may send a GET request to the API of a smart meter vendor to obtain the current reading of the energy consumption of a residential unit. Another example concerns the control of a smart thermostat where the mobile app may send a POST command to the API of the smart thermostat to adjust the temperature. Most of the existing APIs implement a RESTful communication. REST (Representational State Transfer) [30] is an architecture style for designing networked applications. It relies on a stateless, client–server, cacheable communications protocol such as HTTP.

In many cases, a set of APIs and the required documentations are incorporated in a software development kit (SDK). An SDK is a programming package that enables a programmer to develop applications for a specific platform. For example, a billing integration vendor may provide an SDK to an application provider to help the integration with the Customer Information System (CIS) of a utility.

Evolution of Customer-Centric Services

Open Data and APIs solve the problem of accessing and collecting data in a smart grid environment. This is the basis for the development of intelligence and customer-centric applications. Application providers collect, process, and convert raw data to information and useful insights. This service can be delivered to the end user through engaging web and mobile apps. An example is presented in Fig. 4.3. The most significant applications offered by a variety of smart grid service

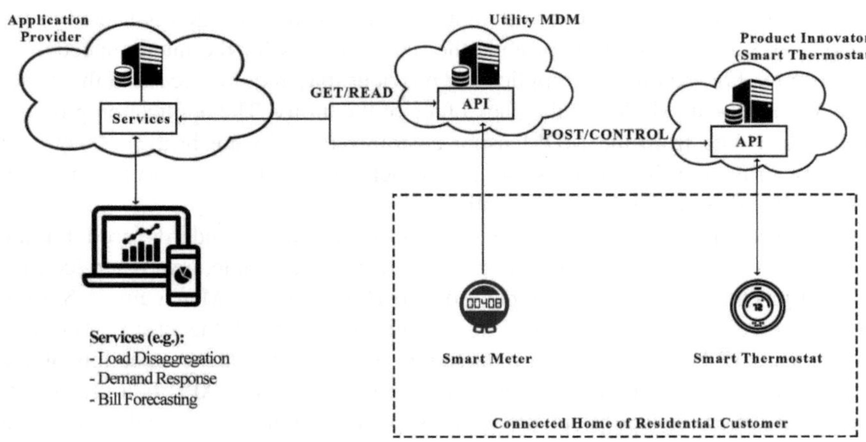

Fig. 4.3 Connection of various parties with web services and APIs for the development of customer-centric applications

providers are those related to energy/bill analytics, energy usage analysis (also known as load disaggregation), demand response, or home energy management and mobile bill payment solutions.

Energy analytics leverage data which is stored in Meter Data Management (MDM) systems. Utilities provide MDM access to third party application developers in a B2B model. In most cases, the application developers create white label applications leveraging MDM data, and the utility offers the applications in a B2C model to the end consumer. Energy analytics applications typically implement features such as bill forecasting, seasonal bill analysis, and daily consumption analysis. These features help end users understand their energy usage and thus optimize their use of energy.

Energy usage analysis (also called "load disaggregation") provides end users with detailed analysis of the energy consumption per appliance. Load disaggregation is more powerful than energy analytics, and is an important service for end consumers due to the detailed insights about appliance consumption. As a consequence, the end consumer can save energy and utilities can create behavioral energy efficiency via programs to engage customers for demand response. Third party application providers may offer services such as load disaggregation in a Business to Business (B2B) fashion to the utility/energy provider who "sells" the service to their customers.

Demand response (DR) and *Home Energy Management* (HEM) focus on load management and automation rather than monitoring. Demand Response (DR) is performed by the utility or the service provider to manage the demand during critical events, such as peak hours. In contrast, Home Energy Management (HEM) provides individuals with the ability to manage the consumption in their homes according to personal preferences. For example, to save money a user may want appliances with high power consumption to operate in off-peak hours when the price of electricity is lower. Gamification is a technique that is widely used to engage consumers in DR and HEM services by providing incentives for participation. For example, a service provider may compare the demand response performance of several neighbors and engage consumers via a community based "competition" to save energy.

Micro Bill Payment solutions leverage the mobile payment preferences of residential customers and provide financial technologies (fintech) and innovation for bill payments. Examples include pre-pay electricity, energy gift cards, and smart energy donations. These types of services fulfill the needs of the sharing economy and the mobile nature of utility customers who are expected to consume energy or share energy outside their meters, e.g., Electric Vehicles. Some new technologies based on blockchain protocol and smart contracts may make possible the development on new digital currencies based on one of the most basic commodities, such as the electricity.

Transactive Grids

Energy in the Sharing Economy

The new energy entities such as rooftop solar panels, electric vehicles (EVs), energy storage, smart thermostats, and smart appliances (aggregated energy management) make the energy resource *demand driven* rather than *generation driven*. The penetration of RES in the power grid, the development of island microgrids, and the evolution of the prosumers create new challenges in the energy sector. The existing regulated and centralized business philosophy and operation leaves a huge pool of resources untapped. Real-time pricing is expected to replace the existing flat pricing architecture to help leverage the time-varying production and consumption at the distribution network. In addition, the mobile nature of electricity, for example, the need of EV owners to consume large amount of energy outside their billing address, and the excess energy of prosumers or VPPs migrates transaction needs from the wholesale to the retail level. There has been already a lot of discussion regarding new concepts emerged from the *fintech* sector such as energy sharing, energy gift cards, and blockchain for energy. This observation is also supported by the fact that sharing economy in smart cities presents an exponential growth.

The concept of Transactive Energy (TE) creates the required foundations for the development of such services and allows consumers, smart appliances, and smart devices to exchange energy related information as well as to transact and interact and be *price-reactive*. These entities are now considered transactive energy agents. Following the definition from [31], TE is a "system of economic and control mechanisms that allows the dynamic balance of supply and demand across the entire electrical infrastructure using a value of a key operation parameter." In the transactive energy grid, the Distribution System Operator (DSO) and the aggregator will play important roles. The Transactive energy concept is expected to help coordinate the "negawatt" market from prosumers and enable the efficient penetration of VPPs in the market that have temporal and geographical dependencies due to distributed energy resources. For example, the production of a VPP changes with time more frequently than a conventional generation facility that is placed in one location and has a seasonal variation. The VPP production may change on an hourly basis since different environmental conditions might occur in various geographical locations.

Transactive Energy Modes and the New Roles

The deployment of highly concentrated Distributed Energy Resources (DERs) and the evolution of prosumers in small geographical regions, such as a Smart City, creates a necessity for a new energy entity, the Distribution System Operator (DSO). Modern utility customers and Smart City citizens have evolved and they

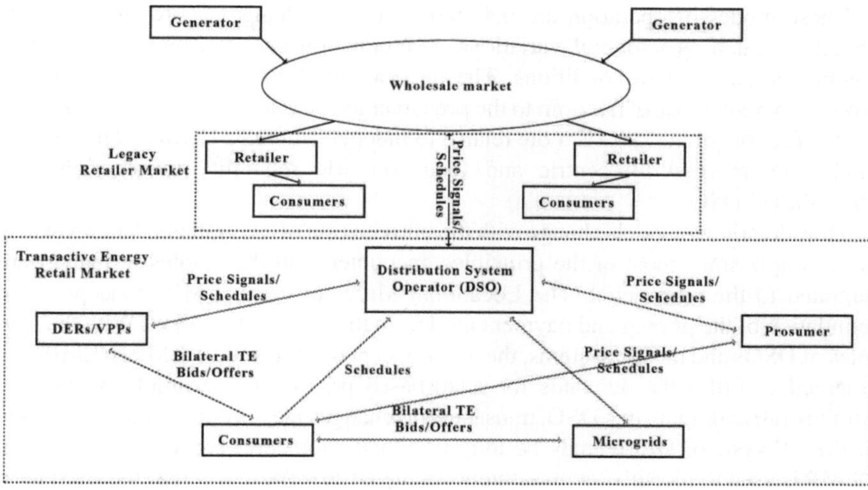

Fig. 4.4 Transactive energy architecture and the role of the DSO

prefer to have more control over their energy use and production. This control takes the form of bilateral transactions. The DSO provides the interface between the environment of DERs and prosumers with the Independent System Operator (ISO). The Regional Transmission Operator (RTO) of the wholesale market protects reliability and efficiency in grid operations and protects the benefits of all entities. An example is presented in Fig. 4.4 [32].

From the legacy hierarchical wholesale market, the TE market provides more degrees of freedom for the active parties to interact with each other. The DSO expands the existing operational role of the distribution utility operator to enable the utilization of DER and demand side management (DSM) as well as leveraging the intelligent transactive devices of prosumers and facilitate peer to peer transactive exchanges. In this distributed and dynamic environment, the DSO facilitates transactions while keeping grid reliability at preferred levels. This is performed by communicating with the wholesale market (ISOs and RTOs) and with the transactive agents. The DSO can be in an independent entity called an IDSO, or it can be managed by the utility.

The TE system incorporates intelligent and transactive grid-edge devices, prosumers, and retail entities (DER, VPP, Microgrid) that may act as transactive agents trying to maximize their own objectives. Managed by the DSO, the following TE system modes can be distinguished [32]:

- "ego-centric" based on local preferences
- Bilateral bid responsive
- Market priced responsive
- Operator dependent.

These modes of operation are not static and may change according to various conditions such as temporal variations of production and consumption, ambient conditions, and system conditions. The first two modes are prosumer-centric and provide more degrees of freedom to the prosumer to act as an independent transactive entity. The role of the DSO is more related to Independent DSO (IDSO). The last two modes are more utility-centric and focus on grid reliability managed by the utility-based DSO.

How do pricing models change with the introduction of TE systems? The answer is quite simple since most of the principles encountered in the wholesale market are migrated to the retail level. The Locational Marginal Price (LMP) concept is the foundation of the pricing and payment models in the wholesale market. With the new roles of DSOs and the TE systems, the new concept of Distribution LMP (DLMP) has emerged to fulfill the demands for retail-based pricing. For example, within the administrative domain of a DSO, transactive exchanges may occur between the agents of the TE system. DLMP may be long term and short term based. The long-term DLMP is used to incentivize investments in the DER market, whereas the short term DLMP is used to provide instantaneous settlements between the TE agents.

Transactive Energy is developed to leverage the growing numbers of DERs and active prosumers of the power grid network by integrating them in a market environment that will achieve greater benefits for all parties. In Smart City environments, this is expected to create new business models that will enhance sustainability and clean energy adoption, and change the role of the end users from passive users to active participants.

Smart Citizens in the Smart Grid

Let us explore what everyday life is in a smart grid/Smart City environment.

Mary is a tech savvy person. She owns a rooftop solar panel, an EV, a smart thermostat and a home battery unit. Her home is smart and connected. She lives in an urban environment which is Smart City enabled. Her smartphone is full of applications related to the health, city operation, environmental monitoring, and social media. She makes almost all of her transactions using her smart phone. The city is within a deregulated energy market and Mary can choose her energy provider. She chooses the one that has the most competitive prices ($/kW) and the one that provides the greatest variety of customer-centric services. She also chooses the provider that is more social and environmentally responsible.

In a typical day, Mary leaves from home to the office in the early morning. Whenever her EV leaves the garage, her smart home goes to sleep mode. The lights switch off and the smart thermostat adjusts to away mode. Her car communicates with the Smart City app and chooses the most eco-friendly route to follow in order to avoid traffic but also high CO_2 concentrations to help Mary with her asthma. Her EV notifies her that a co-worker who is asking for a car pool is on her way to work. She picks him up and they drive together to work minimizing traffic and emissions.

Mary works hard, and her solar panel works hard too. She produces excess energy since her production is larger than her consumption. Her home knows that she is currently away to her office and manages to efficiently allocate the excess of solar energy to charge the home battery sell some portion of energy in the Virtual Power Plant (VPP) of her community as well as donating another portion to a group of low income neighbors. Her clean energy improves people's lives.

During her lunch break, Mary uses the Smart City app and books a restaurant for the night. The Smart City app communicates with the smart home and her smart thermostat is scheduled to have her home comfortable for her arrival. At night she parks her EV at her garage as her Vehicle to Grid charging station monitors the electricity costs and schedules the charging according to minimum prices. During a month, she can save more than 300 kW and 210 kg (465 lb) of CO_2 emissions and make $50 from her energy savings and transactions.

Conclusion

The smart grid changes the role of the consumer and the energy provider and creates a bidirectional relationship of mutual benefit. Utility customers are transformed from passive consumers to active prosumers. At the same time, utilities are transformed from infrastructure owners to enablers and optimizers of new customer-centric services and automation. The same observation is also found in smart cities, where the citizens are transformed from passive absorbers of information to active citizens and enablers of services. The city evolves to an orchestrator of new services and technologies that protects its citizens, the environment, and the economy. Over the coming years, we will experience the integration of the integration of ICT with the energy and other industrial sectors that will create a sustainable, social responsible, and competitive environment.

References

1. www.nrel.com
2. www.census.gov
3. www.eia.gov
4. Harris C (2006) Electricity markets: pricing, structures and economics. Wiley, New York. ISBN: 978-0470011584
5. Pricewatershousecoopers (2016) Looking ahead: future market and business models: future utility business models. PwC report
6. Koutitas G, McClellan S (2017) Smart grid an application development platform. Artech House (in print, expected for publication in September 2017)
7. Carvallo A, Cooper J (2015) The advanced smart grid: edge power driven sustainability, 2nd edn. Artech House. ISBN 978-1-60807-963-6, 2015
8. Accenture (2015) Power play: three new models for growth in the utilities industry. Report. https://www.accenture.com

9. www.smartgridnews.com
10. Lehpamer H (2016) Introduction to power utility communications. Artech House. ISBN: 978-1-63081-006-1
11. Sun Q, Li H, Ma Z, Wang C, Campillo J, Zhang Q, Wallin F, Guo J (2016) A comprehensive review of smart energy meters in intelligent energy networks. IEEE Internet Things J 3(4): 464–479
12. Fettweis GP (2014) The tactile internet: applications and challenges. IEEE Veh Technol Mag 9(1):64–70
13. http://brooklynmicrogrid.com
14. http://www.microgrids.eu/index.php?page=kythnos&id=2
15. Zurborg A (2010) Unlocking customer value: the virtual power plant. U.S. Department of Energy 2010
16. El Bakari K (2014) Smart power systems and markets with virtual power plants: development of distributed energy resources aggregation system. LAP LAMBERT Academic Publishing. ISBN 978-3659566189
17. Lu J, Hossain J (2015) Vehicle-to-grid: linking electric vehicles to the smart grid. The Institution of Engineering and Technology. ISBN 978-1-84919-855-4
18. http://www.sgip.org/
19. National Institute Standards and Techonology. NIST Framework and Roadmap for Smart Grid Interoperability Standards. Release 3.0, 2014
20. https://developer.att.com/blog/smart-cities-with-open-apis
21. https://www.eia.gov/opendata/
22. www.greenbuttondata.org
23. http://www.sgip.org/orange-button/
24. http://pvwatts.nrel.gov/
25. https://developer.enphase.com/
26. https://developers.nest.com/
27. https://www.ecobee.com/developers/
28. https://developer.honeywell.com/
29. Guinard D, Trifa V (2016) Building the web of things: with examples in Node.js and Raspberry Pi. Manning Publications; 1st edn. ISBN: 978-1617292682
30. Fielding RT (2000) "Chapter 5: representational state transfer (REST)" architectural styles and the design of network-based software architectures. (PhD), University of California, Irvine
31. The GridWise Architecture Council (2015) GridWise transactive energy framework. Report, PNNL-22946 Ver1.0. Jan 2015
32. Rahimi F, Ipakchi A, Flecther F (2016) End to end power system operation under the transactive energy paradigm. IEEE Power Energy Mag 14(3):52–62

Chapter 5
The Internet of Things: Nervous System of the Smart City

Yang Zou, Brad Jolly, Ryan Li, Mingyan Wang and Ramandeep Kaur

The Internet of Things

The Internet of Things (IoT) has many definitions, but it generally comprises the set of electronic and photonic devices that communicate over the Internet without human intervention. These devices include some combination of sensors, actuators, and processors, and they typically communicate over the Internet wirelessly. Where the original Internet is for communication between human beings (Web sites, social media, e-mail, mobile apps, VoIP, electronic retailing, etc.), IoT devices communicate without human intervention. If all human beings suddenly disappeared, the data transfer rate on the traditional Internet would quickly tend toward zero, as nobody would be uploading, downloading, or sharing information ranging from scientific data sets to birthday party pictures, music, videos, and other social/business interactions. The IoT would be perfectly capable of running at its current rate of data transfer for a substantial period of time.

Challenges of IoT

The wide range of IoT challenges can be characterized by a single word: scale. Every challenge that exists in today's Internet also pertains to IoT, but the scale is much larger and the impact even more severe. Wireless connected IoT devices will grow exponentially across a wide range of applications. As estimated by Keysight

Y. Zou (✉) · B. Jolly · R. Li · M. Wang
Keysight Technologies, Santa Rosa, CA, USA
e-mail: Yang_Zou@keysight.com

R. Kaur
IOT Solutions, Intel Corporation, Santa Rosa, CA, USA
e-mail: ramandeep.kaur@intel.com

S. McClellan et al. (eds.), *Smart Cities*, DOI 10.1007/978-3-319-59381-4_5

and other industry sources [1], IoT device shipments will reach 10 billion units in 2022. In contrast, shipments of traditional PC, tablet and smart phone will only be 3.5 billion units at that time.

With deployment of large-scale IoT and wireless sensor applications, the amount of data produced will grow significantly. According to IDC statistics and forecast, the global data volume will reach 35 ZB in 2020. As a refresher: 1 ZB = 10^{21} bytes = 1 billion terabytes. The volume of data produced by sensors is 30 times greater than the volume of data produced by human beings [1]. The storage, transmission, and timely processing of massive data will be a challenge.

System Issues

Concurrent device access is another challenge for IoT wireless networks, especially wide area networks. Up to 10000 devices will be accessing one low power wide area network gateway or base station 100 times the typical number of users of a cellular base station!

Existing schedule-based access technologies usually reserve retransmission time slots, and apply frequency division among users to guarantee reliability. These pre-allocated resources are normally not well utilized.

Traditional contention-based access technology has no anti-collision scheme which manages resource utilization. With the increase of data traffic of concurrent applications, network performance will degrade.

Application Requirements

Different applications have different requirements for wireless networks. Consumer applications like smart home and wearable electronics have more requirements on cost, flexibility, and power consumption than on network performance. Smart grid and some other industrial applications have rigorous performance requirements including large scale, low latency, and high throughput. A control application for the smart grid is an example of this requirement: the latency for a small-scale sensor network can be less than 2 mS, whereas the latency for network with thousands of sensors within the substation can be thousands of mS (a few seconds) [2]. Monitoring applications may be less sensitive to price, power consumption, or even latency but have much higher data rate requirements.

There is no single wireless technology which fits the requirements of all applications. The choice of wireless technologies is a trade-off between performance, scale, cost, and power efficiency. Short range wireless technologies, low power wide area network, and even cellular architectures such as "Long-Term Evolution" (LTE) each has a useful place in different IoT applications.

System quality is not exclusively dependent on network technologies and architectures. The quality of single nodes or gateways is also an important factor. Malfunction or under-performance in nodes or gateways may result in retransmissions and in unnecessarily tying up resources. Performance testing of IoT devices and gateways during design and manufacturing is an essential approach to improve overall system efficiency.

Power Consumption

Many IoT devices, especially Smart City or smart agriculture sensors and actuators, have batteries with relatively small capacity. Given the nature of many applications, especially those in remote, hidden, or dangerous locations, customers will often expect a device's battery to last 10 years or more. To respond to this demand, some vendors are producing devices that will run indefinitely, especially if they include some sort of energy harvesting technology.

The choice of battery type depends on the working environment. Safety, low self-discharge, stable voltage, and maximum current are common requirements for IoT batteries. Super-high battery capacity is the ideal case but may not be practical due to cost and size restriction. More often the battery operational lifetime will depend on power consumption optimization.

To optimize battery life for these sorts of demanding applications, modern integrated circuits (ICs) for IoT devices include deep sleep modes that draw current in the nanoamp range. These ICs also include operation modes with slow clocks, LRC circuits that replace crystal oscillators, reduced instruction sets, variable voltages, and other features that reduce current consumption. Characterizing such devices includes measuring current waveforms spanning three to six orders of magnitude in events lasting microseconds or milliseconds.

Figure 5.1 presents a block diagram of a typical IoT gateway system. When a signal from a sensor device enters the gateway system (top left) via a low-power local transmission (e.g., Zigbee), it is processed by the combined Frontend/PA + Radio/SoC module, which is managed by a microcontroller unit (MCU). The Frontend/PA function contains a power amplifier (PA) for outgoing transmissions, and a system-on-chip (SoC) function which performs modulation/demodulation and error management for the received/transmitted signals. The MCU executes software which implements a rudimentary operating system (OS) to manage storage subsystems (Flash), power subsystems (Power + Battery), and other OS-resident functions such as the user interface, networking stack, and onboard devices such as sensors.

Among all components shown in Fig. 5.1, the radio frequency (RF) front-end module, or so-called RF transceiver (TRx), contributes the most to the total power consumption. Sensors and microcontroller unit (MCU) currents are typically a few nano-Amperes (nA) to hundreds of nA in working mode, and down to pico-Amperes (pA) level in sleeping mode. The RF transmitter current ranges from tens of mA to hundreds of mA, depending on output power. The receiver current is lower, normally around 10 mA in Rx mode, and a few hundred nA for register retention.

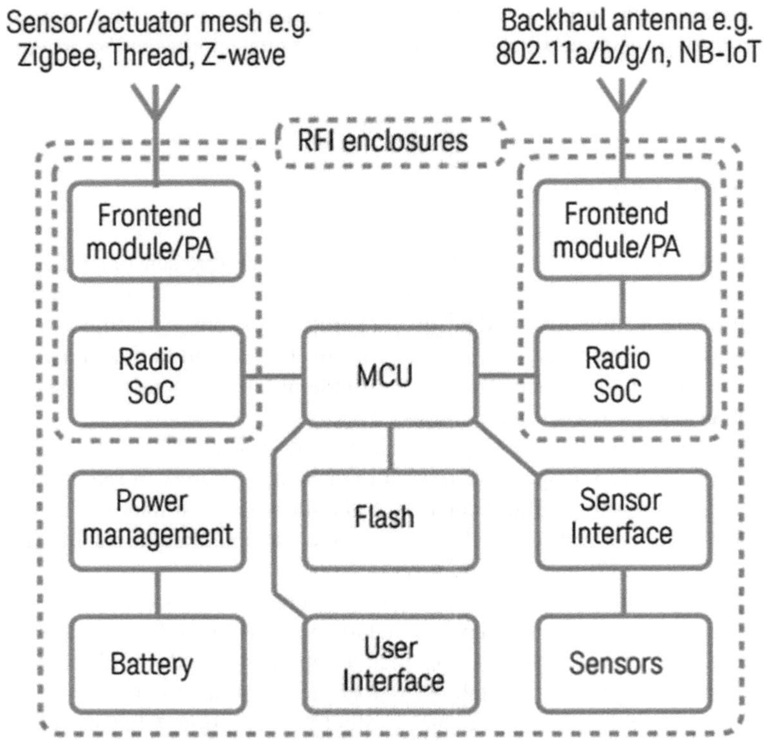

Fig. 5.1 Typical components of an IoT gateway/sensor

Several low power wireless technologies have been defined, such as Bluetooth low energy [3], IEEE 802.15.4 [4], low power WiFi for short range applications, and LoraWAN [5] and NB-IoT [6] for wide area applications. There are three common approaches to reduce power:

1. Power saving mode: nodes will turn off RF TRx when there is no network activity and wake up depending on a chosen scheme.
2. Simple protocol to reduce overhead transmitted over the air. For some IoT applications like automatic meter reading, the data payload is very short and excessive overhead could contribute significantly to transmitter and receiver time and power consumption.
3. Adaptive data rate (ADR) is a method where the actual data rate is adjusted to ensure reliable packet delivery, optimal network performance, and scaling for capacity. With ADR, nodes can choose the highest data rate to reduce the wireless transmission time. Thus, ADR could maximize both the battery life of the node and the overall network capacity.

The product design engineer faces substantial challenges in integrating sensing, processing, control, actuation, and communication components into a product. The designer must understand how the core device and its peripherals behave and consume power. Furthermore, the designer can make firmware choices that control the device's power supplies, analog components, mixed-signal components, digital subsystem, and RF subsystem. In the production test environment, a simpler test set can verify manufacturing quality and device functionality quickly and inexpensively.

The technical diversity of IoT devices requires design engineers to make current measurements with varying levels of resolution and accuracy at each point in the development process. The chip designer must evaluate power consumption in various device operating states very quickly. The module designer needs similar capabilities over a large dynamic range on several chips, and the ability to correlate these with module firmware operations. The product designer may be able to get by with slightly less precise time resolution, but must know the device's overall power consumption throughout the product development process. Finally, the manufacturing test engineer must quickly validate essential device functionality in a time window that keeps up with the speed of the production line.

Sensor Nodes

The growth of the IoT is both driving and being enabled by growth in the sensor industry. The number of sensors produced annually is expected to grow at a compound annual growth rate (CAGR) of about 12% from 2015 to 2021. During the same timeframe, the average selling price is expected to decline from 84 to 74 cents per sensor, but revenue is still expected to grow at a CAGR between 8 and 9 percent [7].

Furthermore, device assembly and testing is a large part of the total cost of manufacturing sensors (60% according to one estimate for accelerometers). Therefore, there is great economic pressure to reduce the need for test and the associated costs [8].

Testing Sensors in IoT Devices

Despite the low cost of IoT sensors, it is important to test sensors accurately for several reasons:

1. Some sensors are used in safety or life-critical applications, such as fire detection systems, automotive airbags, or medical devices. In smart cities, emergency response systems rely on smoke sensors, traffic accident sensors, and other emergency indicators. These sensors must work accurately and reliably.

2. The huge number of sensors deployed annually (some 35 billion in 2016 alone) means that even if 99% of the sensors operate correctly, 350 million will either fail or report inaccurate data. Imagine the havoc and waste a "Smart City" would experience if one percent of its traffic, parking, smoke, and hazardous chemical sensors either failed or delivered wrong information.
3. Some sensors are difficult or very expensive to replace, such as environmental sensors in physically remote locations or medical device sensors inside the human body.

In addition, there is a trend toward incorporating multiple sensors and other microelectromechanical system (MEMS) components in one device. For example, a future smart phone may have a three-axis accelerometer, a three-axis compass, and a three-axis gyroscope (pitch, roll, yaw) in addition to pressure, temperature, and humidity sensors, fingerprint and biochemical sensors, ultraviolet sensors, infrared radiation sensors, and more.

Another factor contributing to the cost of IoT sensor test is the sheer variety of IoT sensors. In addition to the sensors listed above, there are physical event sensors (vibration, tilt, impact, pressure, torque), liquid sensors (presence, level, leakage, flow), and magnetic sensors of various kinds as well as gas sensors that measure presence and concentration of gases both relatively benign (carbon dioxide, oxygen) and dangerous (chlorine, ammonia, carbon monoxide, methane, and volatile organic compounds). Smart agriculture sensors that measure trunk thickness, leaf wetness, soil temperature, and wind velocity, and there are sensors that measure aqueous concentrations of solutes such as iodide, fluoride, chloride, calcium, bromide, perchlorates, and more.

The lists above ignore entire categories of sensors such as power, light, sound, ultrasound, radiation, position, proximity, and presence, and the list seems to grow daily. These sensors also use varying communication protocols, further complicating the sensor test environment, especially in a Smart City context, with its loud and crowded RF communications space. Some of these sensors are very sensitive, and even within a given sensor category, various sensor manufacturers specify their sensors in a variety of ways, posing an additional challenge to R&D engineers and manufacturers who have to figure out how to test to those specifications.

Working Toward Industry-Wide Solutions

To address some of these sensor test challenges, industry groups are working to establish standards, specifications, test procedures, and other best practices for IoT sensor test. For example, the Mobile Industry Processor Interface Alliance (MIPI Alliance) has developed a low-power, two-pin communications specification known as MIPI I3C (Improved Inter Integrated Circuit). This specification defines a bus interface for connecting sensors to an application processor [9].

The Institute of Electrical and Electronics Engineers (IEEE) has produced IEEE 2700-2014, known as the IEEE Standard for Sensor Performance Parameter Definitions [10]. This standard provides common sensor performance specification terminology, units, conditions and limits for accelerometers, gyroscopes, magnetometers, pressure, temperature, and humidity sensors and other sensors. This standard came out of the Standardized Sensor Performance Parameter Definitions (SSPPD) produced by the MEMS Industry Group (MIG).

For example, the IEEE 2700-2014 standard specifies the use and precise definition of these (and other) attributes for accelerometer testing:

- Full scale range.
- Digital bit depth.
- Zero-g offset and temperature coefficient.
- Sensitivity and temperature coefficient.
- Noise.
- Current consumption.
- Output data rate.
- Filter-3 db cutoff frequency.

Successful Testing

To optimize the likelihood of successful test, an IoT sensor product development team should participate in industry groups like the ones mentioned above. This is especially true for newer sensor types. Furthermore, product development teams should leverage the work of these groups in order to avoid the surprises that come with nonstandard approaches to solved problems.

The product managers on IoT sensor product teams should ensure that their products' performance specifications match the end user's application needs. An overspecified sensor adds needless cost for component acquisition, test development, test execution, and rework of perfectly serviceable products. For example, suppose an IoT carbon monoxide sensor for a home or commercial building in a Smart City is specified to work with an accuracy of 1 part per billion (PPB). This would be far beyond what the customer needs, because typical levels in a home range from 500 to 5000 PPB. The typical Smart City user has no useful need to distinguish between, say, 3456 and 3457 PPB, but if the sensor is specified to work to that level, every engineer in the chain (R&D, validation, procurement, manufacturing, and production test) must meet that goal. Even a specification of 10 or 50 PPB accuracy would be more than accurate enough for the actual application.

The product development team should involve manufacturing engineers and manufacturing test engineers in the earliest stages of the product development. These engineers can then make design for manufacturing (DFM) or design for test (DFT) recommendations that will save both time and money if they are incorporated into the product's design.

Testing an IoT sensor or sensor assembly involves two major tasks: providing a known stimulus to the device under test (DUT) and measuring the DUT's response. Providing the stimulus is more difficult than measuring the output for four reasons:

1. The sheer number of DUTs and physical items to be sensed introduces great variety into the challenge.
2. The process by which the stimulus is created and delivered to the DUT must be very repeatable in order to ensure that variations in DUT output are due to sensor variations rather than variations in the physical quantity being sensed or measured. If you are using a high-quality function generator to simulate a physical input, this may be relatively straightforward, but if you are using an actual physical stimulus, such as a gas sample with a known concentration of carbon monoxide, this may be more difficult.
3. The time that a stimulus/DUT combination requires to "settle" at a known value may approach or even exceed the desired beat rate of the production line. For example, suppose the output of a temperature sensor is to be measured at −20, +20, and +60 °C. How long is required for the thermal mass of the sensor to reliably arrive at each temperature?
4. The physical quantity being sensed by the sensor DUT may be dangerous to the user, the equipment, or both, as in the case of a high-temperature sensor, a methane sensor, or a smoke sensor.

In many cases, no special test equipment is needed for the measurement side of sensor tests—a digital multimeter (DMM), oscilloscope, or data acquisition (DAQ) instrument will suffice. However, care should be taken to ensure that the instrument has enough precision and accuracy for process control measurements to quickly detect and eliminate process drift in high-volume production. The R&D, validation, and manufacturing test engineers should also take care to ensure that their probes and lead wires do not add ambient electromagnetic interference (EMI) or uncharacterized parasitic elements into the test path, as sensor signals are often relatively small and therefore subject to noise and measurement error.

Product R&D and validation engineers should work with the module teams that will integrate sensors into IoT modules to optimize sensor integration, including such things as gyroscope mounting and airflow needs. They should also characterize sensor performance across relatively large lot sizes during development and use that data to focus manufacturing test resources on those areas with the smallest margins.

Procurement and manufacturing engineers should take extra care in transporting and storing incoming materials. Many sensors are small and can be damaged by shipping shocks, vibrations, and temperature extremes. Procurement engineers should also work with sensor component suppliers to understand the processes and testing that they use to ensure quality. It may be worth spending a bit more on incoming materials to reduce your own cost of test.

In summary, the growth of the IoT provides both opportunity and challenge to sensor manufacturers and integrators. By participating in the development and adoption of standards, and by considering sensor test goals from product design through manufacturing test, sensor manufacturers and integrators can reduce unnecessary test costs while providing the high-quality IoT sensor products that customers require.

Battery Life

In some cases, it is convenient to run Smart City IoT applications on AC power; the necessary wiring is generally available indoors, and sometimes outdoors as well (e.g., lamp posts). In other cases, AC power is not readily available because the IoT device is either hidden or located where wiring is impractical. For example, a pollution detector may be mounted on the outside of a brick building, and a sensor that detects trash levels in commercial rubbish bins is probably mounted inside the wheeled bin. Providing AC power in such contexts is either expensive, impractical, or impossible, so the customer will require IoT device batteries that last for several years.

Furthermore, the total cost of replacing a battery can be very high, even when the battery is easily accessible. The total costs associated with selecting, purchasing, receiving, stocking, and accounting for a package of inexpensive coin cells may easily double or triple the actual purchase price. Paying for a person to go to the IoT device and replace the battery also adds several multiples to the basic cost of the battery. For these reasons and others, customers often consider battery runtime as a key characteristic when they purchase IoT devices.

Challenges of IoT Battery Drain Analysis (BDA)

Given the desirability of long battery life, detailed and accurate battery drain analysis is an essential part of the product development process. The nature of IoT devices, however, makes this a very challenging task. An IoT device is not like a flashlight, where you can just turn it on with a full battery and measure how long it takes for the light to dim below an acceptable level. IoT devices may take years to run out of battery power, because their average current consumption is often extremely low. Furthermore, the complete current waveform is typically a high bandwidth waveform with a large dynamic range whose exact characteristics depend on numerous design choices made by the firmware engineer who programs the microcontroller unit (MCU). Some key issues in BDA are covered below.

Low-Level Current Measurement

Consider an IoT air quality sensor operated by a coin cell with a capacity of 400 milliamp-hours (mAh) that is specified to have a 10-year battery life. Ten years is slightly under 90,000 h, so the average (mean) current consumption must be approximately 4.5 microamps (μA). Of course, when the device's radio is operating, the device will be consuming hundreds of milliamps, so the time that the device spends in low-power modes (sleep, hibernate, idle, etc.) must consume far less than 4.5 μA. Typically, such modes must consume tens or hundreds of nanoamps (nA). To measure current of, say, 50 nA with 1% precision requires resolution of 500 picoamps (pA) or better, and this can be very challenging for many general purpose measurement instruments.

High Dynamic Range Current Measurement

An RF transceiver with a 600,000:1 dynamic range presents a difficult measurement challenge if the instrument is to avoid the glitching associated with measurement range changes. To measure across a 600,000:1 dynamic range requires 20 bits of resolution in the analog-to-digital converter (ADC) just to cover the range. To get 1% resolution at the low end requires an additional seven bits in the ADC. This type of functionality is very difficult to achieve in general purpose instruments, and instruments with seamless ranging features are therefore popular for BDA in applications with wide dynamic ranges.

High Bandwidth Current Measurements

Smart city IoT devices such as parking lot sensors may sleep for relatively long periods of time (tens of seconds), waking up only for several milliseconds (ms) to perform a quick sensing operation and to perhaps transmit to a base station to indicate a change of occupancy status for a parking space. This whole cycle of wake up (nA to μA), sense (μA to mA), transmit (hundreds of mA), return to sleep (hundreds of mA to nA) occurs in a number of milliseconds, which means the waveform has substantial bandwidth content. The bandwidth of the measurement instrument must substantially exceed the bandwidth of the waveform in order to obtain accurate waveform measurements. Furthermore, it is important to include the bandwidth of the current transducer or probe in addition to the bandwidth of the measurement instrument, as both are important to the overall system bandwidth.

Effects of Firmware Decisions on Battery Life

Because of the highly dynamic nature of IoT device current waveforms, decisions made by the firmware programmer can have a dramatic impact on battery life. Furthermore, these decisions often affect the performance of the device, the amount and accuracy of acquired data, and the user experience. For example, an air quality sensor that averages readings for three seconds will be less noisy than one that takes a single reading over 50 ms. However, it will require 60 times as much energy to make the reading. A water sensor that checks for underground pipe leakage may deliver substantial value if it only wakes up every 5 or 10 min to check for moisture. A firmware programmer for an IoT smoke detector, however, would be well advised to avoid such long sleep periods.

Instruments Used for Battery Drain Analysis

Engineers typically use one or more instruments to perform battery drain analysis (BDA). These instruments include basic instruments found on nearly every bench, such as oscilloscopes and digital multimeters (DMMs). They also may include highly specialized, powerful equipment such as DC power analyzers, precision source/measure units (SMUs), and device current waveform analyzers. The sections that follow refer to these instrument classes generally; the comments are not particular to any particular vendor or model number. The reader is advised to review specifications in the data sheets of individual instruments to determine their suitability for a particular BDA application.

Digital Multimeters

The digital multimeter (DMM) is a relatively inexpensive piece of equipment that is readily available to technicians and engineers. Its simple, familiar use model make it a popular choice for DC current measurements with 4.5–7.5 digits of resolution. Furthermore, DMMs often have the low current measurement capabilities required for IoT devices.

DMMs typically have small, low-resolution displays that make it difficult to see transient signals. They also tend to be low bandwidth instruments, although some high-performance DMMs will have digitization capabilities that have sufficient bandwidth for many IoT applications. They also tend to lack the sophisticated analysis features of more specialized BDA instruments.

Oscilloscopes and Current Probes

Like the DMM, an oscilloscope is a relatively inexpensive piece of equipment, with a familiar, simple use model. Unlike the DMM, however, the oscilloscope has a large screen that allows the user to view, pan, and zoom in on the waveform to find and examine transient elements of the signal in detail. In addition, the oscilloscope typically has many math, measurement, and triggering functions that provide great control over measurements and let users get to answers quickly.

Oscilloscopes generally have bandwidth in the hundreds of MHz or GHz range, but they do not have the low-level current accuracy and resolution of a DMM. They also take up more space on a desktop or in a test rack than a DMM typically does.

DC Power Analyzers

A DC power analyzer can be an excellent tool for BDA, especially if it includes a source/measure unit (SMU) that can simulate battery characteristics. A good DC power analyzer may provide seamless ranging, which allows you to cover the wide dynamic range from nA to mA without the glitching associated with range changing. A DC power analyzer usually will have more measurement and triggering capabilities than a DMM, but not as many as an oscilloscope. Furthermore, DC power analyzers often have software applications that run on a PC and provide BDA-specific capabilities that are beyond those provided by oscilloscopes.

Another advantage of a DC power analyzer is that the user can develop a test using an SMU module on the bench, and then use the same SMU module in manufacturing test. This improves time to market and gives the design engineer greater control and confidence over production test. The main downside of DC power analyzers is that they are not as ubiquitous as oscilloscopes or DMMs, and they may be more expensive. Their bandwidth also typically falls between that of the DMM and the oscilloscope.

Precision Source/Measure Units

A precision source/measure unit (SMU) is of limited use for BDA, but it can be the ideal choice for low-level current measurements. Precision SMUs have the best accuracy and resolution for these types of measurements, but they have limited bandwidth, small displays, and limited triggering and analysis features.

Device Current Waveform Analyzers

Device current waveform analyzers are a new class of instrument with excellent capabilities for BDA. They have the large display and powerful math, measurement, and triggering functions of an oscilloscope, but they have much better current measurement accuracy and resolution than an oscilloscope. A device current waveform analyzer also lets the user quickly view, pan, and zoom in on the waveform, just as an oscilloscope does. A device current waveform analyzer also has built-in software features that give fast and detailed answers to challenging BDA questions.

The primary downside of device current waveform analyzers is cost. Unlike a DMM or oscilloscope, they are too expensive to put on every engineer's bench, but their capabilities make them an important tool to include in every workgroup.

Software Tools for BDA

Many instruments will have associated software tools that perform BDA functions. These may be built into the instrument, or they may be external applications that run on PCs or tablet computers and connect to the instrument to obtain the measurement data. Some instruments also have control software that allows you to remotely control a virtual bench of geographically dispersed instruments from a single PC.

Complementary Current Distribution Function (CCDF)

A CCDF is an X–Y chart that shows what portion of the battery's power is consumed at various current levels. A representative CCDF, produced by a Keysight CX3300 Series Device Current Waveform Analyzer, is shown in Fig. 5.2. The horizontal axis shows various current levels, and the vertical axis is a percentage of battery charge consumption that starts at 100% and goes down to 0% as charge is consumed at various current levels. Note that the CCDF always goes from 100 to 0%; the full range equals the amount of charge consumed in the test, which is not necessarily the full runtime of the battery. For example, in Fig. 5.2, just 18% of the charge is consumed at currents above 4 mA.

Automatic Current Profile

An automatic current profile provides a detailed graph and table that indicates how much of the battery's total charge is consumed in each segment of a device's current waveform. The segments are typically automatically identified by the profile

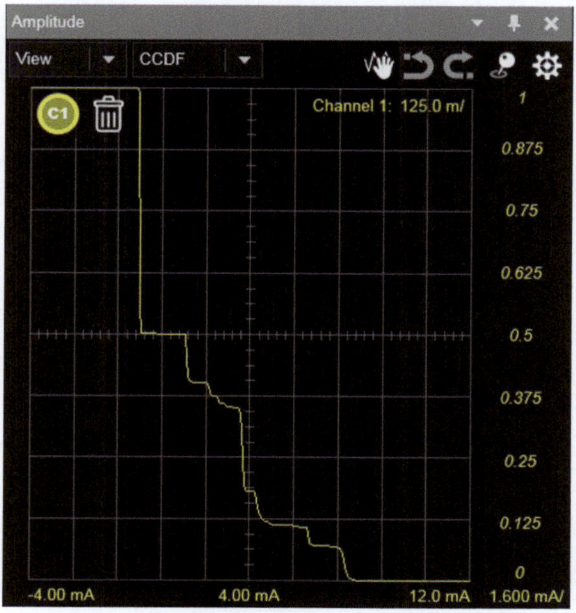

Fig. 5.2 Complementary current distribution function (CCDF)

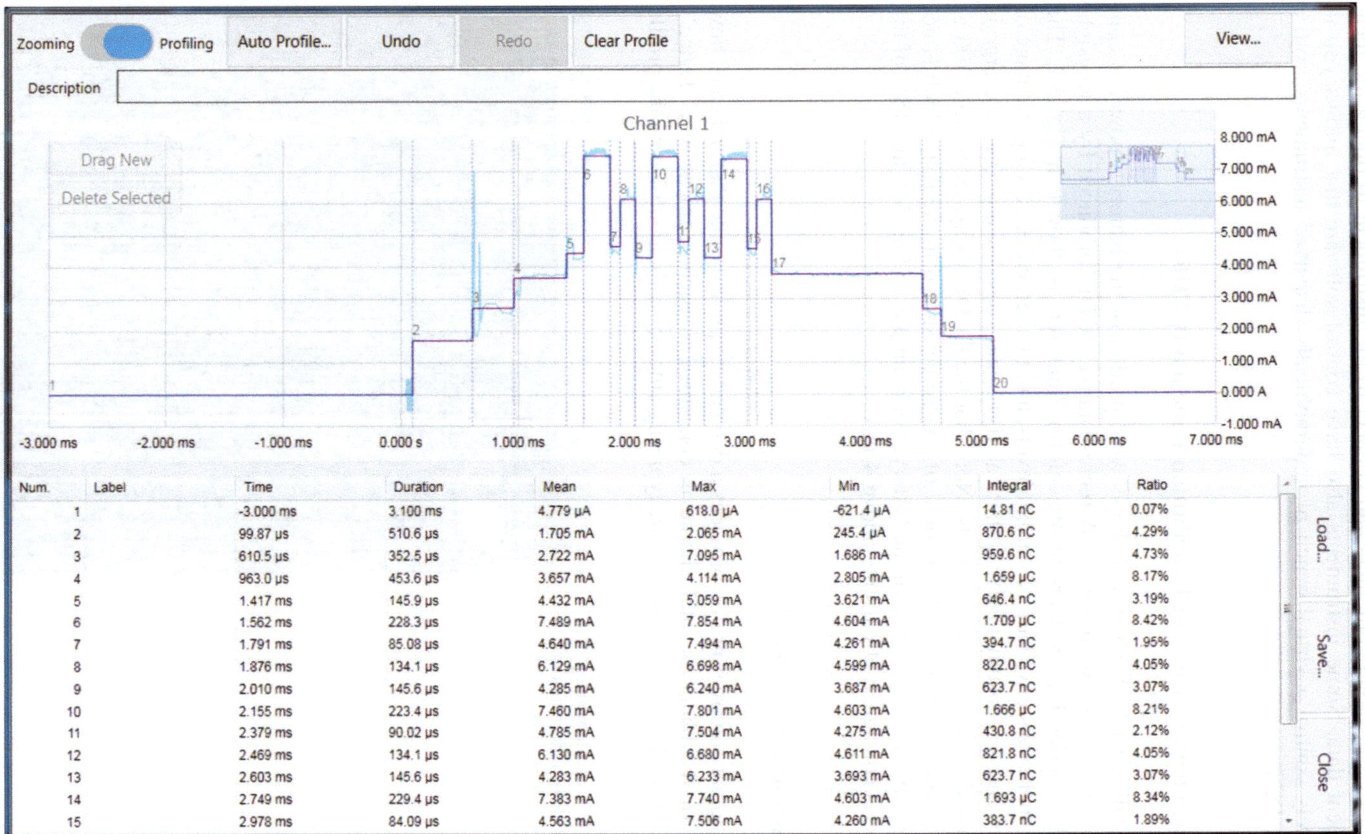

Fig. 5.3 Automatic current profile (ACP)

software, and the user can adjust the parameters used for this automatic identification. The user can also add, delete, or move segment boundaries to focus on specific areas of interest that may differ from the segments that were automatically identified by the software. An automatic current profile produced by a Keysight CX3300 Series Device Current Waveform Analyzer is shown in Fig. 5.3.

Regardless of how the profile segments are determined, the software typically provides the following information for each segment in the current waveform profile:

- Start time.
- Duration.
- Minimum current.
- Average current.
- Maximum current.
- Charge consumed.
- Percentage of total charge consumed.

Additional Considerations

Of course, there is much more to optimizing battery runtime than simply using high-quality BDA instruments and software. For example, in order to ensure repeatable measurement results, you need to make sure that the batteries you use in BDA are of the same age and stored and operated at consistent temperature. You also need to optimize the architecture of your device, pay close attention to the MCU firmware programming, and consider using energy micro-harvesting to recharge your battery.

Temperature Considerations

A 3-week-old battery that has been sitting loose in an office drawer may perform differently than a 10-month-old battery that has been stored in its original packaging in an unheated warehouse. A test performed in a Finnish R&D lab at 18 °C may yield different results than a similar test performed in a Malaysian lab at 25 °C. Furthermore, the same outdoor air quality sensor may consume energy at vastly different rates depending on whether it is deployed in Oulu or Penang.

Architectural Optimization

The architecture of your device will directly affect its battery life. For example, some microcontroller units (MCUs) include efficient DC–DC buck converters that allow you to specify voltage levels for MCUs and peripherals that can operate across a range of voltages. A transceiver that can operate at voltages ranging from 2.2 to 3.6 V will have substantially different power consumption, output power, attenuation, and blocking characteristics at different input voltages, so flexible DC–DC conversion is a significant advantage. Also, an MCU may integrate RF radio capabilities, sensors, and other peripherals. Such integration may save current by offering greater control over optional and configurable features on these peripherals.

MCU Firmware Programming

There are several steps you can take in programming firmware to extend battery life. Write your program to finish each task quickly and to sleep for as long as possible. Cycle power on sensors and other peripherals so that they are on only when in use. Of course, when you cycle sensor power, consider the power on stabilization time of the sensor to avoid unduly affecting measurement accuracy.

If possible, use features in your peripheral modules to replace functions that would otherwise have to be executed by the MCU. For example, timer peripheral modules may be able to automatically generate pulse width modulation (PWM) and receive external timing signals.

Use all of the typical good programming practices, such as setting constants outside of loops, using macros carefully, avoiding declaring unnecessary variables, unrolling small loops, shifting bits to replace certain integer math operations, and similar techniques. Also, turn on all compiler optimizations to clean up any optimizations that you might have missed.

Energy Harvesting

Energy harvesting capabilities can dramatically extend the lifetime of IoT devices. In some cases, they can extend the device's battery life; in other cases, they can eliminate the device's battery altogether either by charging a capacitor or feeding power into a power management IC. Typical energy harvesting technologies include solar microcells, piezoelectric components, and thermoelectric devices that harvest energy from ambient light, vibration, or temperature differentials.

Narrowband IoT: Overview and Test Challenges

NB-IoT (Narrowband Internet of Things), the new narrowband radio technology developed for the IoT, is one of the key low power wide area network (LPWAN) technologies based on the cellular telecommunication licensed spectrum. It is standardized by the 3rd Generation Partnership Project (3GPP). The standardization of NB-IoT core in 3GPP Release 13 was completed in June 2016. Three technologies defined by 3GPP, including NB-IoT, may be useful in addressing different application requirements in the IoT market.

The objectives of the NB-IoT solution are:

- Extensive deployment flexibility (Software upgrade based on existing cellular infrastructure).
- Improved indoor coverage (164 dB link budget which is +20 dB compared to GSM).
- Ultra-low cost devices (targeting less than $5 per module).
- Low power consumption (10+ years battery life).
- A massive number of devices of low data rate (50 K devices per cell with small data rates).

The NB-IoT target application categories include smart metering for water, gas, etc., smart agriculture, health care, pets and asset tracking, and many more.

Overview and Key Parameters

NB-IoT evolved from 3GPP's Long Term Evolution (LTE), but it is a new radio access technology. It inherits many aspects of the LTE-like frequency bands, physical layer foundation, and the reuse of higher layer architecture, which have all been optimized to meet the NB-IoT solution goals.

NB-IoT uses the same frequency space as LTE, and a subset of LTE frequency-division duplex (FDD) bands are defined for NB-IoT. In 3GPP Release 13, only the FDD duplex mode is defined for NB-IoT, but it requires forward compatibility with time-division duplex channels (TDD). The NB-IoT key parameters are shown in Table 5.1.

Table 5.1 NB-IoT key parameters

Parameter	Value
Frequency range	LTE FDD bands: 1, 2, 3, 5, 8, 12, 13, 17, 18, 19, 20, 26, 28, 66
Deployment mode	Stand-alone, in-band and guard-band
Duplex mode	FDD half duplex type B
Bandwidth	200 kHz (1PRB)
Multiple access	Downlink: OFDMA Uplink: SC-FDMA
Modulation schemes	Downlink: BPSK, QPSK Uplink: Single tone: $\pi/4$-QPSK, $\pi/2$-BPSK Multi tone: QPSK
MIMO	No MIMO support
Coverage	164 dB (+20 dB GPRS)
Latency	<10 s
Mobility	Nomadic (only reselections, no handovers)

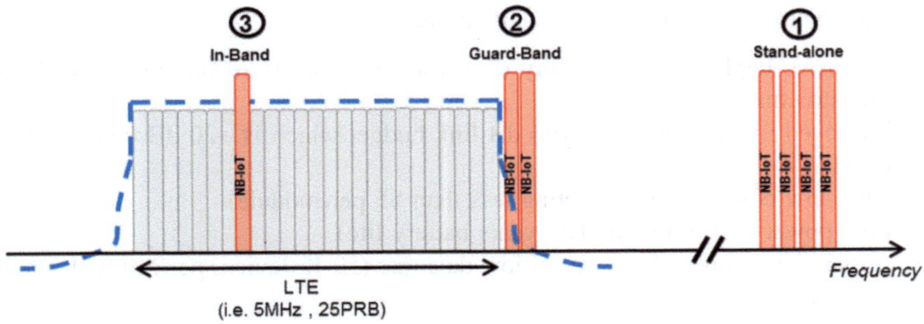

Fig. 5.4 NB-IoT spectral allocation

Three deployment modes (Stand-alone, Guard-band and In-band) have been defined for NB-IoT, and are shown in Fig. 5.4, which is taken from 3GPP Work Item [11].

1. Stand-alone: replacing a GSM carrier with an NB-IoT cell.
2. Guard band: utilizing the unused resource blocks within a LTE carrier's guard band with guaranteed co-existence.
3. In band: using part of an LTE carrier.

Design and Test Challenges

NB-IoT as a new narrowband technology brings up test challenges to achieve NB-IoT objectives and target applications requirements, such as improved indoor coverage, ultra-low device cost, and low device power consumption. Design and test of NB-IoT devices are faced with numerous challenges. Included among these challenges are substantial issues related to reliability, coverage, and battery life.

Reliability

Many IoT devices are expected to be deployed in the field and function without any human assistance for many years. As such, they need to support software updates and handle unexpected situations such as server breakdown, memory leaks and similar issues. An NB-IoT device needs to be significantly more robust than a smartphone, for example, given that no direct human intervention is available.

Coverage

IoT devices can be deployed in difficult to reach places, e.g., underground water well, and still need to guarantee good coverage. NB-IoT requires a 164 dB link budget and defines three different coverage levels in the standard. The antenna effects, multi-radio/inter-mode interference, and many other factors could influence the coverage, all of which will require thorough test and validation.

Battery Life

IoT devices require battery life of more than 10 years. In order to meet that goal, the NB-IoT standards define an enhanced Discontinuous Reception (eDRX) and a power saving mode (PSM). Testing battery drain in different network conditions is key to guarantee a successful deployment and meet the expected quality of service.

Feature Enhancements

The NB-IoT Release 13 is currently frozen and enhancements are being considered for Release 14. Some of these are discussed below.

Positioning Technology

Positioning technology is very important for several important NB-IoT applications. Asset tracking is an example of such an application. Positioning technologies known as "Observed Time Difference of Arrival" (O-TDOA) and "Uplink Time Difference of Arrival" (U-TDOA) are particularly useful in NB-IoT systems. Both O-TDOA and U-TDOA are based on principles of multilateration, where the "time difference of arrival" (TDOA) of a signal is measured from groups of synchronized receivers or transmitters.

To implement O-TDOA, a mobile system must measure the time of arrival for signals received from three or more synchronized transmitters. An O-TDOA measurement scenario is shown in Fig. 5.5. In the figure, the mobile device is labeled "UE" corresponding with the 3GPP standard reference for "User Equipment," and the three base stations are labeled "eNB," corresponding with the 3GPP standard reference for an "Evolved NodeB" system in an LTE Radio Access Network. The UE measures the time of arrival (t1, t2, t3) from the three base stations, and computes time differences (t2-t1) and (t3-t1) based on the reference signal from its "Serving Cell" (eNB1).

Using these two sets of time differences, the UE is able to compute its position relative to the base stations, subject to some uncertainty. The uncertainty inherent in the time-of-arrival measurements results in location uncertainty for the UE calculations, which is shown in Fig. 5.5 as the small gray area around the UE.

U-TDOA is supported under some conditions in the current standardization of so-called "5G" mobile communications systems, which began in 2016 with 3GPP Release 14 [12]. A measurement scenario for U-TDOA is shown in Fig. 5.6. In contrast to the O-TDOA scenario, three base stations (eNB) measure the time of

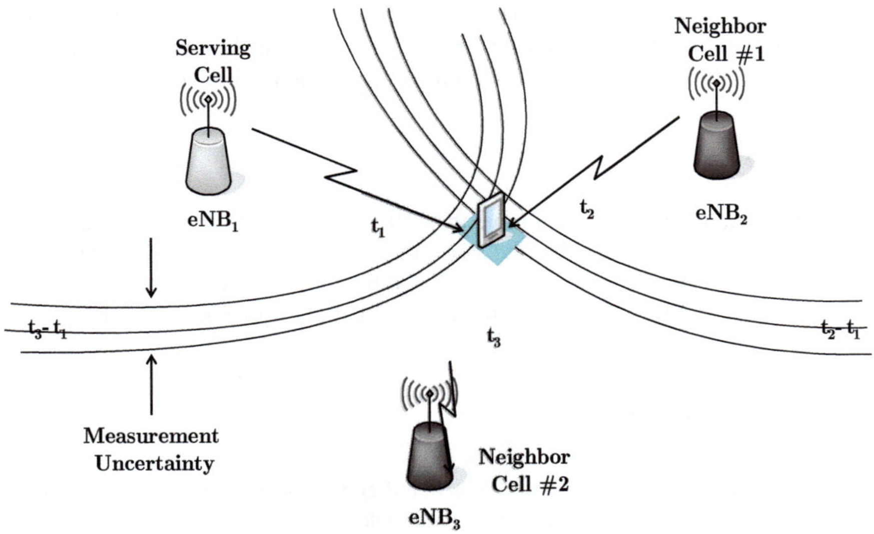

Fig. 5.5 Observed time difference of arrival (O-TDOA). *Source* Clip art courtesy of VRT Systems (http://www.vrt.com.au/downloads/vrt-network-equipment)

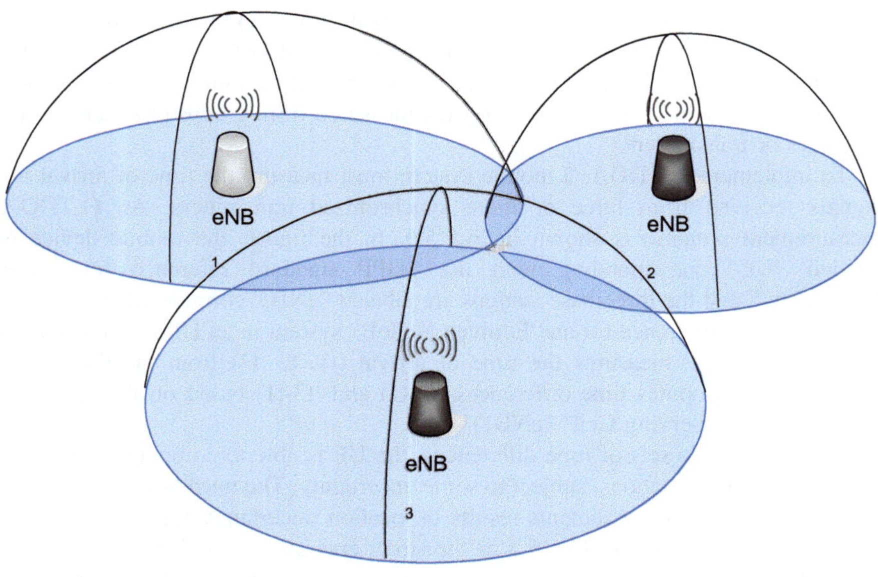

Fig. 5.6 Uplink time difference of arrival (U-TDOA). *Source* Clip art courtesy of VRT Systems (http://www.vrt.com.au/downloads/vrt-network-equipment)

arrival for a signal transmitted by the mobile device (UE) and estimate the UE's position based on measured time differences.

Each of these technologies has advantages and disadvantages related to important IoT concepts, such as positioning accuracy, complexity, and power consumption.

Other Enhancements

A number of other enhancements are being considered in 3GPP Release 14 and beyond which affect NB-IoT applications. A few of these enhancements are described briefly below.

- Enhanced SC-PTM: Single-cell point-to-multipoint technology achieves NB-IoT downlink multicast application requirements, such as sending firmware, software, and group messages.
- Multiple-carrier enhancement: Data transmission on non-anchor carrier to improve the whole system capacity and performance.
- Mobility and service continuity enhancements: Cell-reselection is defined in release 13 for NB-IoT. Mobility enhancement to improve service continuity without power consumption increasing will be considered in Release 14.
- Power consumption and latency reduction.
- New Power Classes: In release 13, UE power class 3 and class 5 are defined. New classes will be evaluated in release 14, for example, a new UE power class with lower maximum transmit power that is suitable for small form factor batteries (e.g., wearables).

Conclusion

This chapter has discussed several aspects of Internet of Things (IoT) technologies which are important in the development of smart cities. These aspects include the design, challenges, requirements, test methods, and standards which affect IoT devices and systems. Unique challenges of IoT systems are presented, including considerations surrounding battery life and testing, sensor design and testing, and wireless network parameters and considerations.

References

1. Feldman S (2012) Unified information access: creating information synergy. IDC
2. Internet of Things: Wireless Sensor Network (2014) International electrotechnical commission (IEC) wireless sensor networks project team. http://www.iec.ch/whitepaper/internetofthings

3. Bluetooth Specification, Version 4.2, Dec 2014. https://www.bluetooth.com/specifications/adopted-specifications
4. IEEE Standard for Low-Rate Wireless Networks, IEEE Std. 802.15.4-2015, Apr 2016, ISBN:978-1-5044-0845-5
5. LoRaWAN 1.0.2 Specification. https://www.lora-alliance.org
6. Standardization of NB-IOT completed, 3GPP, 2016 http://www.3gpp.org/news-events/3gpp-news/1785-nb_iot_complete
7. Lightman K (2017) Welcome and introduction, MEMS and sensors industry group track, CES 2017. https://memsindustrygroup.site-ym.com/?ces2017resources
8. Precision Testing for MEMS Accelerometers. Science and Technology Research News, April 2016. http://www.scienceandtechnologyresearchnews.com/precision-testing-mems-accelerometers
9. Gaitan M (2015) Survey of practices used for accelerometer performance parameters in datasheets. International Electronics Manufacturing Initiative (iNEMI) end of project report, Sep 2015. http://thor.inemi.org/webdownload/2015/EOP_MEMS_081515.pdf
10. IEEE Standard for Sensor Performance Parameter Definitions, IEEE Std. 2700-2014, Aug 2014, ISBN:978-0-7381-9253-6
11. RP-151621, 3GPP Work Item Description "Narrowband IOT," 3GPP TSG RAN meeting #69, Phoenix, USA, Sep 14–16, 2015
12. Parkvall S (2015) Release 14—the start of 5G standardization, Ericsson Research Blog, 2015. https://www.ericsson.com/research-blog/lte/release-14-the-start-of-5g-standardization/

Chapter 6
The Cloud: A Critical Smart City Asset

Brad Booth

Introduction

The Digital Revolution brought about the Information Age. Humankind has advanced the ability to collect and manipulate digital information as technology has progressed. In a relatively short period of time, technology has advanced so dramatically that society can create information and share that information with anyone on the planet. One of the key elements in enabling this growth of information is the ability to communicate over vast distances at almost the speed of light. The Internet, a vast collection of networks, provides the ability to move information from one person to another virtually uninhibited. As the Internet came into being, the Information Age accelerated with the desire to communicate with family and friends, to search "online" for information, and to conduct business.

This advancement of technology since then has led to the development of the ability to go almost anywhere and remain connected whether via laptop, tablet, or smart phone. It is now possible to take a transoceanic flight while remaining connected to the Internet for the entire journey. It enables the ability for driverless cars to travel our city streets.

The "Cloud" has existed from the moment the Internet was made available to the world. It is, in its simplest form, a location to store, manipulate and retrieve data; a data warehouse or a data center. What has changed from the simple data center of the past to become a Cloud data center is the type of information stored. In the early days of the Internet, if you wanted to have an email address you either had to create your own email server at home or purchase the service from a provider. When search engines and browsers started to come into being, companies started to build data centers to provide their employees with email access or to permit their customers to find their business offerings.

B. Booth (✉)
Microsoft, Redmond, WA 98052, USA
e-mail: bjbooth@gmail.com

S. McClellan et al. (eds.), *Smart Cities*, DOI 10.1007/978-3-319-59381-4_6

The Cloud became a consolidation of data centers. The Cloud companies provided other companies the ability to have a data center without having to build one or staff it with information technology (IT) professionals. Suddenly a Ma and Pa shop located anywhere could have a global presence with a website, catalog, email address, and the ability to accept payment. People and companies no longer had to worry about where their data was stored because the Cloud companies created a large enough presence to provide virtually instantaneous access to their information.

This pervasive, global presence of your data somewhere out there on the Internet is what led to the coining of the term "the Cloud."

Architecture of the Cloud

The basic architecture for a Cloud data center is not that different than an Enterprise (corporate, government, university, etc.) data center. Servers store, manipulate, and retrieve information. A network connects all these servers to each other and to the Internet. The big difference is the scale of the Cloud datacenter. Scale not only in size and the number of servers, but also locations on a global scale.

Because the Cloud data center has such a large concentration of information and data from individuals, corporations, governments, universities, etc., there are some key requirements the Cloud providers focus on:

(1) Security
(2) Scalability
(3) Reliability
(4) Performance
(5) Power.

In the following sections, each of these requirements will be covered in more detail.

Security

One does not have to look too far to see the number of companies that have had security breaches [1]. For this reason, security is paramount in a Cloud data center. Because Cloud data centers contain information from more than one company or one person, there are often multiple attempts to either obtain information or to prevent access to the information (denial of service—DoS) by hackers. A DoS attack is where the hacker or perpetrator seeks to make a machine or network resource unavailable to its intended users by temporarily or indefinitely disrupting services of a host connected to the Internet. DoS is typically accomplished by flooding the targeted machine or resource with superfluous requests to overload systems and prevent some or all legitimate requests from being fulfilled.

Cloud data center providers need to be diligent because of the volume of attacks that occur. They are a focal point due to the volume of information that they contain. It is for this reason that Cloud data center providers have cybercrime divisions [2]. Security is paramount to the operation and protect of their customers' information.

Data and information security is also affected by geography too. Many countries have rules and regulations about data contained within or transitioning through their country. For example, data being back up from one region or country may have to transition through another country, and the Cloud data center provider needs to understand whether or not the data can leave the region or country, and the rules and requirements that may impact that data in transition.

Data has two states: data at rest (stored on a server) and data in transit (moving through the network). Data at rest is protected by security protocols inside the data centers. The facilities often have perimeter fencing, biometric sensors, and security protocols that must be followed to gain access to the data at rest. Data at rest can also be encrypted to prevent those gaining electronic access from reading the information.

Data in transit also has levels of protection. Unlike data at rest which is contained within the data center, data in transit can be moving both inside and outside the data center. Outside the data center is more vulnerable due to the inability to use the same security measures for a data center facility.

In these instances, Cloud data center providers often use one or more security protocol. For example, SSL (secure sockets layer) encryption is often used by software or web sites to protect sensitive data. Encryption can also be done by hardware using protocols such as MACSec (Media Access Control Security). Where SSL offers protection across the whole link from the application running on the customer's device to the cloud server, MACSec offers security on each physical link between network elements.

Scalability

Scalability of a Cloud data center has two distinct aspects: expansion and refresh. Expansion is the ability to grow existing or build new data centers. The term data warehouse was using in the early days of Cloud data centers because warehouse was a perfect description of the facilities used to house these data centers. Like warehouses used by shipping companies or retailers, the Cloud warehouse-scale data centers were typically built where land was affordable. There were two other elements that were critical to building these massive facilities: connection to the Internet and power.

There are many places in the world where land and power are readily available, but without a connection to the Internet, there is no ability to be a Cloud data center. It is for this reason, that you will often find very specific areas where multiple Cloud providers will build their data centers.

Fig. 6.1 Microsoft datacenter

Cloud data centers are large-scale facilities. Figure 6.1 shows the scale and size of an existing Microsoft data center.

This facility is an excellent example how Microsoft dealt with being able to expand and refresh their Cloud data center. The large warehouse building at the top was their existing data center and to rapidly expand the facility Microsoft developed and deployed container-based servers known as pods [3]. Pods can be seen on the right-hand side and in the lower left of Fig. 6.1.

Whether a Cloud provider uses a warehouse or pods, the goal is the same; to be able to expand and refresh rapidly. Expansion can happen either at an existing facility, like in the image above, or by building new facilities. Refresh is what occurs with existing equipment.

Refresh in a Cloud data center is the ability to quickly replace existing servers and networking equipment whether it is in a warehouse or in a pod. This refresh is sometimes referred to as a crop rotation. Because the facility will exist longer than the equipment used inside, Cloud providers have to update the equipment on a cadence that keeps the equipment competitive relative to performance and power usage.

To be able to scale to meet the growing demand for Cloud services, Cloud providers often will be expanding and refreshing a facility at the same time.

Reliability

One of the key metrics that users of Cloud services, whether they are individuals or corporations, is reliability. Reliability is designed into a Cloud architecture. There are a few elements that help with reliability:

(1) Replication
(2) Redundancy
(3) Distribution.

Replication applies to both data that is stored and data that is being manipulated [4]. For data that is stored, Cloud providers will often permit the end user to decide where their information is stored. There are typically three options: local, regional (or zone), and global (or geo).

Local replication stores the end user's data in multiple locations inside a specific data center. Each replica resides in a separate fault and upgrade domain to prevent loss of information. A fault domain is a group of servers that represent a physical unit of failure; whereas, an upgrade domain is a larger group of servers that will be impacted during a refresh.

Local replication may still be desirable in certain scenarios:

- Fast replication,
- Application stored data that can be easily reconstructed, or
- Data governance requirements (ability to store data in foreign countries).

Regional replication stores the end user's data across multiple datacenters within multiple regions, rather than within a single data center. This replication operates like local replication by creating multiple copies of the data in multiple data centers within a region. Regional replication offers higher reliability than local replication if the primary datacenter is unavailable. The data replication is slower due to the distributed nature of the replication.

Global replication stores data to another region that is hundreds or thousands of miles away from the primary region. This form of replication provides access to the data if there is a complete regional outage or a disaster in the primary region. Like local and regional replication, multiple copies of the data are created in multiple data centers in multiple regions. Global replication may not meet the data governance requirements.

Redundancy works in an equivalent manner as replication by making sure there are machines and networks available for the application to use. This is commonly referred to as the "nines" of availability. The nines refers to a desired percentage of availability of a given computer system or data center. The term five 9s refers to 99.999% availability.

High availability is achieved by eliminating single points of failure, reliable distribution of workloads and detection of failures as they occur. The single points of failure can be any part of the data center where the application is running. If that application is not distributed to other servers in the data center, then the application has a single point of failure. Replication of data across a broad set of machines in a data center, region, or geography greatly improves the ability for the application to run on multiple servers with access to the information it requires.

Another potential point of failure is the power to the servers. Cloud data centers are designed with large battery systems to support the time it takes for the backup generators to start after a power failure. The battery systems prevent the slight power glitch that typically occurs between power failure and backup generator initiation.

This same principle is applied to application running in the data center. For data that is being manipulated or being used in an application, that operation typically is distributed. To prevent operations from occurring all on one machine, Cloud data center providers use load balancers. Load balancers are a way to spread requests or operations over a broader set of machines. Instead of an application waiting for a specific machine to respond, the application can respond to the customer based upon the first machine to respond.

By distribution the workload over a data center designed with redundancy where the data is always available due to replication, Cloud data centers are able to provide five 9s of reliability.

Performance

Performance is a difficult metric to use in Cloud data centers. While many Cloud data centers have similar architectures, today they have very different target markets. For example, Google is well known for their search engine and Gmail system, Microsoft for its Office tool set, Facebook for social media, and Amazon for shopping. A customer using a Cloud provider may do so for specific reasons or specific applications. It is because of this, that performance metrics are only used where Cloud data centers overlap in a competitive landscape.

That being said, Cloud data centers focus on performance of their systems not only to provide a competitive advantage, but also because as technology advances the cost per bit, whether stored or computed, improves.

Technology ages. In the cell phone market, a phone is almost out of date by the time it hits the shelves. While that trend has slowed slightly, it was common for a newer version to appear on the market every year. The same happens in Cloud data centers. There is an ability to amortize the cost of the equipment used to build the data center. If one was to use a 3-year amortization period, that would equal the historical trend for CPU improvements, known as the Intel Tick-Tock Model [5]. Each tick and tock results in increased performance and energy efficiency while also adding new capabilities. As a CPU ages past a tick-tock cycle (typically within 3–4 years), inefficiencies increase. This is particularly true when running the latest software and applications. This results in a less-than-competitive solution in performance and energy consumption.

Power

Power is the primary operating expense of a Cloud data center. With rows upon rows of servers and networking equipment, electricity is what keeps the data center running. Once the equipment is deployed in a Cloud data center, other than some maintenance and refreshing of equipment, the power consumed by a data center is the majority of the cost of ownership; hence, the desire for Cloud data centers to be extremely conscious of the how the power is consumed, but also how the power is generated and the cost of that power.

For example, US data centers consume between 70 and 100 billion kilowatt hours (kWh) of electricity per year [6]. To give a sense of the scale, the average US home consumes about 11,000 kWh per year; therefore, US data centers consume power equivalent to about 7–9 million homes.

That figure is for all US data centers, not just for Cloud data centers, but it gives the sense of the power required to run the Internet. As was mentioned earlier, where Cloud data centers differ from typical data centers is the scale at which they operate. Cloud data centers want to compete not only in the services they offer their customers but also the energy efficiency of their data centers.

The Green Grid (thegreengrid.org) was founded in 2007 with the collective viewpoint that energy efficiency in the data center was one of the most significant issues facing technology providers and their customers. The organization developed specifications that permitted data center providers to calculate the energy efficiency of their facilities. The power usage effectiveness (PUE) was created to provide a ratio of the total amount of energy used by a data center facility to the energy delivered to the computing equipment. Anything that is not considered computing equipment falls into the category of facility energy. For example, lighting and cooling are power used by the facility, not by the computing equipment.

$$PUE = \frac{Total\,Facility\,Energy\,Consumption}{Total\,Computing\,Equipment\,Energy\,Consumption} \tag{1}$$

While PUE can be an effective tool to perform comparisons, there is some basic knowledge that is required when using the tool. For example, a facility operating in a cooler climate is likely to have a lower PUE than a facility operating in hotter region. There is also the matter of what is included in the total facility energy consumption [7]. The PUE can be calculated based upon what the Cloud operator feels is part of their facility.

Whether the PUE is calculated with inefficiencies like transformer loss or parasitic energy drain, the power consumed by the Cloud data centers has approached being within 10–15% of the optimal value. This is vast improvement over the average data center PUE of about 1.8.

As Cloud data centers became more energy efficient, many of the Cloud operators investigated and implemented alternative sources of affordable power. One of the cheapest forms of renewable energy is hydroelectricity. It is not hard to understand why there are so many data center facilities built in locations where hydroelectricity is cheap and plentiful [8, 9]. Cloud data centers also use other alternatives sources like wind [10], solar [11], or biogas [12].

All sources of power, whether conventional or alternative, are impacted by inefficiencies. While PUE can be a tool to perform comparisons, PUE does not account for the power lost in transformers, rectifiers and voltage regulators, or the energy consuming in moving electrons or photons, or the electro-optical translation. Power is also consumed by equipment fans, backup battery charging systems, control systems, and more.

For example, a retimer is an integrated circuit that reconditions an electrical signal to reset the system's jitter budget. A single retimer consumes roughly 500 mWs of power. If retimers were used in a Cloud hyper-scale data center, the total power consumed by those simple little components could use up to 1 MW of power. While a retimer may not draw much power on its own, the scale of a Cloud data center can result in a very measurable impact.

Cloud data center providers are aware of this level of impact; therefore, they focus on understanding how power is consumed, and maximizing productive uses while minimizing waste.

Conclusion

This chapter has discussed the origins and structure of "the Cloud" as well as its role in the emerging "Smart City" phenomenon. Cloud data centers provide focused, highly secure, and reliable services for their end customers. These facilities have performance and power metrics that are superior to enterprise data centers while also providing customers with the flexibility to scale and grow as their storage and computing demands change. The "Smart City" of the future will depend on increasingly data-intensive technologies and processes. As a result, access to "the Cloud" will become a critical requirement for increasingly distributed, personal, autonomous, and "smart" services.

References

1. The Huffington Post (2017) Credit card hack. http://www.huffingtonpost.com/news/credit-card-hack. Accessed 10 Mar 2017
2. Warnick J (2015) Digital detectives: inside microsoft's new headquarters for the fight against cybercrime. https://news.microsoft.com/stories/cybercrime/index.html. Accessed 16 Feb 2017
3. Miller R (2009) Microsoft unveils its container-powered cloud. http://www.datacenterknowledge.com/archives/2009/09/30/microsoft-unveils-its-container-powered-cloud. Accessed 22 Mar 2017
4. Macy M, et al (2017) Azure storage replication. https://docs.microsoft.com/en-us/azure/storage/storage-redundancy. Accessed 25 Mar 2017

5. Intel Corporation (2015) Intel tick-tock model: the tick-tock model through the years. http://www.intel.com/content/www/us/en/silicon-innovations/intel-tick-tock-model-general.html. Accessed 23 Mar 2017
6. Shehabi A, et al (2016) United States data center energy usage report (LBNL Report Number LBNL-1005775). https://eta.lbl.gov/publications/united-states-data-center-energy. Accessed 29 Mar 2017
7. Higginbotham S (2012) Whose data centers are more efficient? Facebook's or Google's? https://gigaom.com/2012/03/26/whose-data-centers-are-more-efficient-facebooks-or-googles. Accessed 30 Mar 2017
8. Kennett S (2013) Facebook's hydro-electric powered data center goes online. https://www.2degreesnetwork.com/groups/2degrees-community/resources/facebooks-hydro-electric-powered-data-center-goes-online. Accessed 30 Mar 2017
9. Rogoway M (2013) Google contemplates new data centers in The Dalles. http://www.oregonlive.com/silicon-forest/index.ssf/2013/04/google_contemplates_new_data_c.html. Accessed 30 Mar 2017
10. Data Center Knowledge (2009) Wind-powered data centers. http://www.datacenterknowledge.com/wind-powered-data-centers. Accessed 30 Mar 2017
11. Data Center Knowledge (2009) Solar-powered data centers. http://www.datacenterknowledge.com/solar-powered-data-centers. Accessed 30 Mar 2017
12. Verge J (2014) Microsoft opens zero-carbon methane-powered data center in wyoming. http://www.datacenterknowledge.com/archives/2014/11/07/microsoft-opens-zero-carbon-methane-powered-data-center-wyoming. Accessed 30 Mar 2017

Part III
Transportation Considerations

Chapter 7
Transportation Electrification

Karl Popham

Introduction

Any Smart City strategy that wishes to address clean air and climate change has to consider emissions from the nation's two largest sources; transportation and electricity generation. A smart transportation electrification strategy can concurrently reduce both sources of emissions while providing other community benefits of affordability, energy security, and electric grid reliability to include maximizing the viability and value of renewable energy sources.

This chapter will discuss both the benefits and real-world solutions deployed since 2011 by the nation's 8th largest public electric utility serving Austin, Texas.

BENEFITS: The Smart City Business Case for Transportation Electrification

A smart transportation electrification strategy is well aligned with a larger Smart Cities vision as it offers real-world and scalable solutions to improve air quality, combat climate change, and support a city's affordability goals. Also, transportation electrifications promote "fuel independence" by replacing oil imports with locally generated electric fuel. Additionally, Electric Vehicles (EVs), with its inherent energy storage capabilities, has a strong synergy to support an increased penetration of distributed PV solar generation.

K. Popham (✉)
Austin Energy, Austin, TX 78704, USA
e-mail: Karl.Popham@austinenergy.com

S. McClellan et al. (eds.), *Smart Cities*, DOI 10.1007/978-3-319-59381-4_7

Cleaner Air and Climate Protection

The majority of US greenhouse gas emissions come from two primary sectors; electricity and transportation (see Fig. 7.1). Transportation electrification addresses both of these emission sources and is a critical component of Austin's vision of realizing Net-Zero emissions by 2050 [1]. The basic strategy is to displace the burning of petroleum with 100% renewable energy for transportation.

Electric Vehicles produce zero tailpipe emissions and research shows they also produce less total carbon dioxide (CO_2) emissions compared to petroleum-fueled vehicles regardless of where in the U.S. they are physically plugged [2]. Additionally, with an ongoing trend of increased electricity generation from cleaner and more renewable energy sources, the reduced emissions benefit continues to grow. Austin's electric utility, Austin Energy, has a renewable energy goal of 55% by 2025. Austin Energy's public charging network, Plug-In Everywhere™, is backed by 100% renewable energy, specifically its GreenChoice® renewable energy credit program powered by Texas wind. Furthermore, the city has a vision to achieve a community-wide net-zero emissions goal by 2050. Such a vision requires strong, innovative, and leading programs to address the emissions from both transportation and electricity sectors to be successful. Clean transportation is a critical component of any serious plan to address climate change.

Affordability

In 2011, EVs were seen as mostly toys for the rich as demonstrated by the Tesla with its high-performance, luxury cars priced $80,000 on up and out of reach for most buyers and definitely not a product for Low-to-Moderate Income (LMI) communities. Since then however, EVs have seen a significant price drop

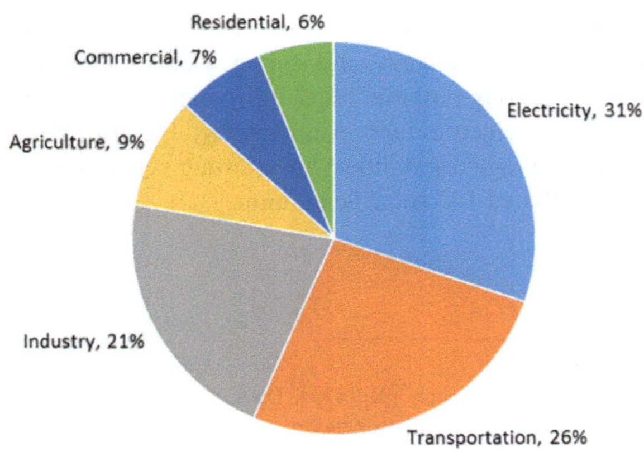

Fig. 7.1 National US greenhouse gas inventory (2015) by sector [7]

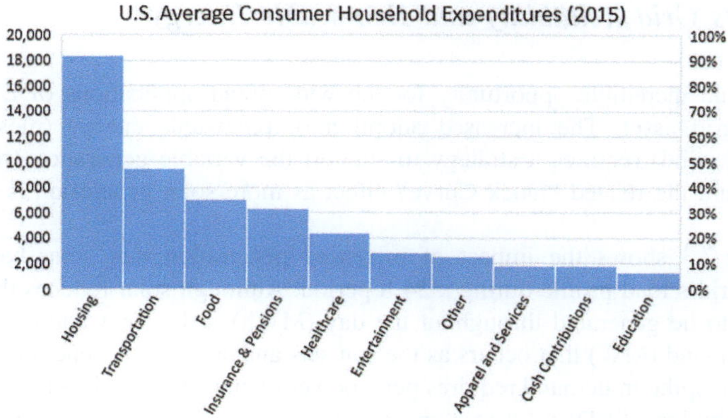

Fig. 7.2 US household expenditures by category (2015) [8]

Table 7.1 US petroleum imports and exports 2015 (million barrels per day) [10]

Import sources	Gross imports	Exports	Imports
Total, all countries	9.45	4.74	4.71
OPEC countries	2.89 (31%)	0.24	2.65
Persian Gulf countries	1.51 (16%)	0.02	1.49

due to the reduction of lithium ion batteries, economies of scale, and increase manufacturer competition. In 2017, a resident of Austin can purchase a new EV for $13,500 (after $7500 federal incentives) and the EV used-car market is just now arriving as first generation EVs go off leases making it feasible to purchase a low mileage EV for $6000–$8000. This compares very favorably with the average cost of a new mid-size car at $25,095 [3].

Reducing transportation costs can be a significant source of cost savings for the average household. As shown in Fig. 7.2, in the US, the second largest household expense is transportation and as such products and services that significantly reduce this expense can make an impactful change to consumers.

The primary use of oil in the United States is for transportation. As shown in Table 1, in 2015, the United States imported 9.45 million barrels of petroleum per day with 47% of this import from Persian Gulf and Organization of the Petroleum Exporting (OPEC) Countries. On the other hand, electricity is almost exclusively derived from domestic resources with an ever increasing mix from renewable sources like wind and solar. EVs can contribute significantly to supporting domestically produced energy while reducing transportation cost market risk as electricity is historically a very stable cost of energy especially when compared to petroleum due to its volatility from geopolitical factors and natural disasters [4].

Supports Grid Reliability and Renewable Energy

There is an incredible opportunity for the widespread applications of EVs as an electric grid asset. The increased adoption of renewable energy (mostly from solar and wind) requires a strategy to support the variable generation from these sources and the related "Duck Curve" effect as more solar generation is added to the grid.

Figure 7.3 shows the impact of increased penetration that solar has on the electric grid's load profile during a 24 h period. Although solar reduces the energy required to be generated throughout the day (MWh) it does not address the peak power demand (MW) that occurs as the sun sets and solar stops generating energy. This sharp spike in demand requires peak power plants and associated transmission and distribution (T&D) infrastructure to deliver energy, typically from natural gas generation. Distributed Energy Resource (DER) management, to include utilizing batteries has great potential to reliably transition power during this curve thus reducing overall cost and emissions from this effect. Battery storage also has applications in grid resiliency, power quality, market ancillary services, and more.

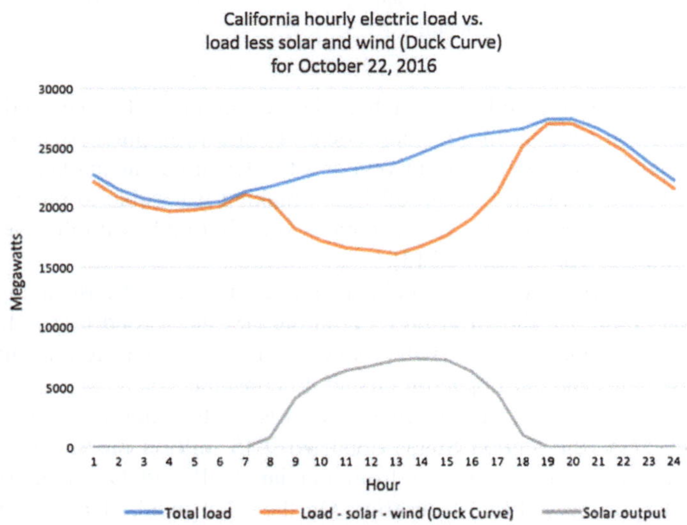

Fig. 7.3 California 24 h electric load demonstrating the "Duck Curve" from increased solar adoption [9]

SOLUTIONS: Smart City Transportation Electrification Projects and Programs

As discussed previously, there is a strong case for transportation electrification as critical components of a Smart City roadmap. But what does this mean exactly? What does a potential portfolio of projects look like? What are the lessons learned from rolling out many of these projects in Austin and what is the strategy moving forward?

Providing Public Charging Stations

Typically, one of the first items a community looks at in being "EV Ready" is to provide public charging infrastructure. EV charging stations are summarized in Table 2, along with cost, throughput, and application types. A Smart City EV fueling strategy requires understanding the different types of charging speeds, standards, costs, and potential impact to the electric grid. Most Electric Vehicle Service Equipment (EVSE) manufactures push for Level 2 and DCFast solutions as they offer the highest miles per hour of charge but are also the most expensive stations to deploy and provision. As such, understanding how low-cost level 1 fits into a charging infrastructure portfolio is also important.

Level 2: For Level 2 infrastructure, Austin Energy's growth from 113 charging ports in 2012 to over 500 (and growing) in 2017 is almost solely from collaborating with businesses, employers, and apartment communities (e.g. "hosts") by providing rebates and turn-key operations of the charging stations at a host's site. The host provides the physical space, initial purchase, and installation of the station. Austin Energy rebates up to 50% of the upfront cost, manages the station, charges customers to use the station, and refunds back to the host the electricity consumed by the station if installed behind their meter.

DCFast: DCFast is typically used to top off vehicles or support highway travel. For DCFast the Level 2 rebate model is not effective due to the high upfront capital costs compared to Level 2. As such, the utility is deploying its owned and operated stations. Unlike Level 2 charging with a universal standard, DCFast connections to car manufactures have opposing charging standards, CHAdeMO for Asian market

Table 7.2 EV Charging Station Types

Type	Demand (kW)	Range (mi) per hour	Capital cost per station	Best application
DCFast	50+	200+	$80,000–$130,000	Corridor traffic, specific vehicle types (e.g. Bus)
Level 2	6	24	$3000–$8000	Public, retail, workplace
Level 1	1	4	$70–$90	Home, multifamily

Fig. 7.4 Austin's "Electric Drive" DCFast dual-standard station

cars, SAE Combo for US and European cars, and Tesla's Supercharger standard. Recently, single stations that support all major standards have become available. This new multi-standard is the configuration Austin Energy is rolling out for its owned and operated stations to include its high-visibility launch of Electric Drive. Austin's DCFast charging station is shown in Fig. 7.4.

Level 1: Often overlooked in charging infrastructure rollouts, Level 1 is a critical component of any overall EV charging strategy and approximately half of all charging done in Austin is from level 1 outlets. A level 1 is a typical (and in many cases existing) standard 110v outlet. It provides for around 4 miles of charge per hour and its main advantages is the ubiquitous availability of level 1 outlets and its low cost to deploy more, if needed. Many customers find this very convenient, not unlike the daily habit of charging a phone. And from a utility perspective this is also advantageous as it still generates kWh (revenue) via an overnight "trickle-charge" when electric rates are at their cheapest while utilizing existing electric service delivery infrastructure.

Lessons Learned: It is important to take a strategic and diversified charging infrastructure approach to ensure you are using the most cost-effective solution per application. This includes today's Level 1, Level 2, and building your DCFast infrastructure with multi-standard support and with the groundwork for even faster charging equipment in the near future (100 kW +). It is also important to have a creative financial model to remain competitive to baseline competition, a gallon of gas.

Ensure Affordable Electric Fuel Costs

In Austin, the average residential kWh is about 10 cents so a plug-in electric vehicle (PEV) or battery-powered electric vehicle (BEV) will get about 35–40 miles of range per dollar from plugging into an existing home wall outlet. In Austin, those that wish to fuel up on over 500 charging ports at workplace, multifamily, retail, and public spaces can get unlimited fill-ups for a flat rate of $4.17/month. Austin Energy also released a pilot program, called EV360, for unlimited off-peak charging at home and unlimited charging 24 × 7 on its public network for a flat fee of $30/month.

Lessons Learned: For decades utilities have focused on volumetric pricing (kWh) when in reality the majority of its costs are fixed. Not unlike the transformation of telco companies moving from pay-per minute plans to a much more robust service plans, utilities are well positioned to move into more innovative pricing models with transportation electrification programs. For example, Austin Energy launched a flat $4.17 per month pricing for public fueling. Pricing models are profitable and scalable for a utility when you look at a customer behavior as a group and not focus on a single, "worse-case" customer. Also, utilities need to consider new revenue and value from customer's that chose to plug at home under existing rates. From examining the customer base, typical miles driven, and rates it is estimated every EV in the Austin market is worth on average $400/EV/year of new revenue to the local utility. Having competitive and customer-friendly service plans helps stimulate demand.

Launch an Outreach/Marketing Campaign

Austin's phase 1 approach was to first promote the idea of driving electric cars. Called the "Charge Forth" campaign, it looked very unlike a traditional utility marketing campaign. It was a very clean design, no "fine print", no costing information but simply photography and a tag-line to promote the concept of driving electric. Recently, the city's utility has launched its second campaign, "The Ultimate Conversation Starter" and "electricity > gas" to further expand the idea of driving electric and now with more specific programs, EV calculators, rebate information, and testimonials on tangible benefits to the customer. This information is readily accessible at its program's website, www.PluginAustin.com.

Lessons Learned: A key, especially for utilities, is to position both campaigns as a creative led campaign. Many utilities approach marketing with a wide range of stakeholders to contribute content and visual input into the campaign. The end result with so many "cooks in the kitchen" is often bland materials that mimic a coupon. Putting your creative professionals in the center results in a much more attention grabbing campaign. Another key is focus on the factors people truly look at in buying a vehicle. What internal research validated is cost and environmental concerns, although a factor, were not top of the list for most car buyers. Instead, having a car that is fun to drive and a reflection of their lifestyle is on top of that list and campaigns should focus accordingly.

Partner with Auto Dealerships and Manufacturers

It is a basic idea but it is important to remember car companies make and dealerships sell cars (an obvious exception is the Tesla direct model that does both). Car buying is a personal choice and for many an emotional buy. Active outreach to local car dealerships and meetings with car manufacturers help design programs and a strategy for both organizations. The utility's most popular dealership program was to sell directly to the dealership unlimited, one-year public fueling cards for $50 each. The utility also sent staff to help train dealership sales teams on EV basics and utility programs.

Lessons Learned: First find a good car dealership partner who is committed to selling EVs (do not assume that every dealer that has EV availability is this way). Develop programs that work for them. Then, when calling on other dealerships always start with the number of units being sold by this "lighthouse" dealer and how you can partner with them to support similar results. Also, the dealership pre-pay fueling card was a great way to impact the car buyer at the point of sale with another compelling reason to make the switch to electric.

Integrate Electric Vehicles into the Grid

There are two fundamental ways to manage EVs as an asset to the grid; managing the one-way charge of EVs or by managing a two-way flow of energy, often referred to as Vehicle to Grid (V2G). Managing charge (typically referred to as Demand Response) is relatively straight forward and gets the majority of the value to the electric grid while mitigating the major barriers of V2G. V2G holds exceptional promise for grid reliability, resiliency, and ancillary services but still has a few years to evolve as a saleable solution.

The energy storage available on most EVs is significant. A current generation LEAF is 30 kWh, the BMW i3 is 33 kWh, and the Tesla ranges from 60 to 90 kWh. Comparatively, a residential energy storage system, the Tesla Power Wall, is under 7 kWh. So any strategy on distributed energy storage should include the opportunity of community-wide EV adoption.

EVs as a Demand Response Asset: Austin Energy and with funding from a federal Department of Energy ARPA-E grant, conducted a pilot to treat vehicle charging as a Demand Response (DR) asset in the summer of 2013. Using open standards, the utility was able to successfully interrupt charging during peak demand events thus reducing electricity cost, reduce grid operator transmission charges, and reduce peak demand. The lesson learned is that open standards, and in this case OpenADR, was a much easier and integrated approach improving its viability. The alternative API approach was cumbersome and problematic. Another way to shave peak production is via a Time of Use (TOU) rate. Austin Energy launched a "home and away" program, called EV360, that allows for unlimited off-peak charging and 24 $\times$ 7 public charging for a flat rate of \$30/h.

EVs as Vehicle to Grid (V2G) Asset: The other fundamental way to manage EVs as an asset is by managing a two-way flow of energy where the energy stored in EVs supply power back to the home or the electric grid. The major barriers that exist today to this approach is a general lack of car manufacturer support for this application, lack of UL listed devices in the US to transfer the energy back from the car to the home/grid, and typically a non-favorable business case unless the customer is in a very high-cost electricity market. Austin is leading a US Department of Energy Project called Austin SHINES that integrates PV Solar and energy storage at the grid, commercial, and residential scale. Although using stationary storage, the open-standards approach can be applied to EV adoption.

Lessons Learned: There is tangible value today in incorporating EVs into a Demand Response program using open standards like OpenADR. V2G applications in the US will be more readily available/scalable starting in 2020 as there are still several hurdles to address.

Support EV Adoption of Transportation Network Companies (TNC) and Taxi Fleets

Taxis and Ride-hailing companies (like Ride Austin and Lyft) are a great opportunity for electrification. Not only are these typically high-mileage vehicles to maximize clean air/emission benefits they are also high-visibility vehicles exposing potential new EV drivers to the experience of driving electric. Currently, Austin Energy is deploying its new DCFast rollout with a criteria to include service for these vehicles. Also, the utility is partnering with stakeholders to develop a TNC hub in Austin that would allow a fleet of EVs to be leased by Austin drivers to provide new vehicles, a great ride, and lower Total Cost of Ownership compared to a driver buying or leasing a similar gas vehicle. This business model also helps address "ladders of opportunity" by providing vehicles to those that may not be able to afford their own car yet still want to be a TNC driver.

Support Autonomous Vehicles and Their Electrification

In 2017, Austin City Council passed a resolution to "shift the City's transportation system to one that enables 'Shared, Electric, and Autonomous Mobility Services' and provide a City ready and willing for it." Austin is no stranger to Autonomous Vehicles (AV) as Google has been testing its cars for years on Austin's public streets but what makes this resolution particularly interesting is the linking of AV to EV.

Having these two linked not only addresses the city's mobility needs into the twenty-first century but by linking with transportation electrification also takes advantage of a disruptive technology to realize the city's vision of cleaner air and protection from climate change as outlined in its Net-Zero Emissions by 2050 plan. Additionally, EVs with their large batteries are also well suited to power the complex electronics and navigation systems of AVs.

On average, 95% of a car's life is spent parked, making it a highly underutilized transportation asset [5]. The business model of autonomous ride-hailing has the promise of greatly reduced costs by increasing the utilization of this asset while removing the operational costs of a driver. Additionally, AVs are well positioned to combine the conveniences of ride-hailing apps (e.g., on-demand, door-to-door service) while greatly undercutting costs and inconveniences of car ownership (e.g., fueling, maintenance, parking, etc.).

AV Public Buses and Shuttles: Autonomous Vehicles also looks to transform public transit by significantly reducing its operational expense of the driver while improving safety. Additionally, electric buses are quicker, relatively quiet, and have zero tailpipe emissions. Through lowering costs, improving services, and using electric AV shuttles for first-last mile mobility solutions there is an opportunity to greatly increase ridership and revenue while at the same time reducing costs.

Electrify Public Fleets

On May 5, 2016 Austin City Council passed a resolution directing staff to conduct "an assessment to determine the benefits, timeline, and feasibility of increasing electric vehicle adoption into the City's Fleet Services vehicles." Austin Energy, City Sustainability Office, City Fleet Services, and the Electrification Coalition developed a plan to include favorable cost savings associated with a larger rollout of electric vehicles. The plan developed calls for the natural retirement of 330 petroleum-fueled vehicles with EVs (72 Plug-in Hybrids, 258 Battery Electric Vehicles) over 3.5 years with projected 10 year TCO savings to the city of $3.5 M [6].

Lessons Learned: Not having readily available plug-in electric SUVs and trucks is a major restriction on electrification but this gap is being addressed by manufacturers. Also, Austin's plan excluded police vehicles (that make up over 625 vehicles in Austin) although that barrier is being addressed and a good example is a 100 vehicle Police BMW i3 rollout in Los Angeles, California. Also, the analysis from fleet services is that buying EVs is now the same or even a bit lower than comparable gas-powered fleet vehicles, removing increased upfront costs.

Provide EV-Related Consumer Incentives/Rebates

Austin Energy provides up to 50% hardware and installation rebates for residential, retail, workplace, and multifamily charging stations. This combined with a turn-key operation has seen its public station growth go from 113 in 2012 to over 500 charging ports today. The utility also have a low-cost but high-visibility program to rebate electric bikes and scooters. The key reason for residential level 2 rebates was to establish a contractual relationship, called "Plug-In Partners", with early EV adopters. This gives the utility a go-to list of pilot participants to include a pilot group for the ARPA-E grant-funded residential EV demand response pilot in 2013.

Develop a Multifamily EV Program

Over 40% of Austin's population lives in multifamily residences and having a program to reach out to this demographic is essential. As such, Austin Energy launched a specific multifamily campaign that included an 80% rebate with unique branding for host properties. Specifically, this program was for apartment rental communities as multifamily condos with Home Owner Associations (HOAs) have completely different barriers and mostly revolve around not wanting to subsidize anything to one particular owner. But by focusing on apartment rental communities, this resulted in Austin Energy deploying EV stations to 37 apartment communities that are home to approximately 21,000 residents. And although Level 1 is a great application at apartment communities, most early adopters wanted the visibility and amenity of Level 2 stations.

Lessons Learned: Apartment management companies are very competitive in the amenity business and there is a "domino" affect that happens once you get your first few properties onboard. Highly promoting early adopters generated a lot of interest caused from other apartment communities that did not want to get left behind in offering this amenity.

Establish a Public Space for Electrification

Having a public space where the community can come together to demonstrate electric vehicles is an important part of a community outreach campaign. In March 2017, Austin Energy formally launched via a ribbon cutting ceremony with the Austin Mayor, Austin Energy GM/CEO, and other leaders the opening of Electric Drive. Components of this project include a dual-standard DCFast charging station, Level 2 charging station, name designation to Electric Drive, Solar-powered kiosk with integrated battery storage to charge electric bikes and scooters, bike share station, and outreach displays. This kiosk is shown in Fig. 7.5. Having a formal launch event helps drive excitement from the community and media to demonstrate Austin is EV ready.

Electrify Public Transit/Buses

Austin Energy is working with its local public transit, CapMetro, to provide a public transportation energy rate to minimize electric fuel costs. This uses a Time of Use (TOU) strategy to promote electric fueling to be done off peak (typically

Fig. 7.5 Austin's "Electric Drive" solar charging kiosk

avoiding 3 pm–7 pm). To support environmental benefits, all electricity from this new fueling program will be backed by 100% renewable energy (West Texas wind credits) through Austin Energy's GreenChoice program.

Go After Grants/External Funding Sources

Since 2012 Austin Energy has participated in numerous EV related grants to provide resources and focus to rollout public infrastructure, reduce barriers to adoption, and pilot emerging technologies to integrate EVs onto the electric grid. There is a lot of federal, state, and philanthropic funding resources readily available to provide resources to advance a community EV program. Specific grant-funded projects (past and present) at Austin's utility include:

- ChargePoint America—US Department of Energy (DOE) grant that installed the first 113 public charging stations in Austin.
- Central Texas Fuel Independence Project—DOE grant provided training for first responders, a job training program at the local community college, bus electrification analysis, and fleet electrification tools.
- Texas River Cities EV Initiative—DOE grant provided a detailed (500 + pages of content), community roadmap to implement electric vehicle infrastructure.
- Demand Response Pilot—DOE grant successfully demonstrated open standards and feasibility for EV residential demand response programs.
- Pecan Street Research Project—Austin Energy is one of six original co-founders and 100% of the financial match to this DOE grant that launched this innovative research organization that includes several EV pilots.
- Austin SHINES project—Although not an EV specific grant, this DOE funded project is launching PV Solar with storage integration to include grid, commercial, and residential batteries. By using open standards, this project can readily be adopted to any electric storage asset to include electric vehicles.
- Low-Income EV Project—11th Hour Project grant, a philanthropic organization, is working with Austin to develop EV programs specifically for low-income and multifamily residents.

Lessons Learned: When writing a grant, make sure you bring in partners that demonstrate you have the expertise and plan in place to execute on time and within budget. Align your project objectives with specific metrics of the grant. Establish a good track record and go after grants and projects that are in your core competency of success. Being a Principal Investigator on most of the above projects, does make me appreciate the level of reporting required to successfully manage a grant. Specifically, from highest administrative burden/documentation to least I would rate the most task intensive as being Federal Grants with "Substantial Involvement" clause, Federal Grants in general, Texas state grants, and the easiest to administer are philanthropic grants.

Conclusion

As discussed, transportation electrification can play a critical role into realizing the goals of a Smart Cities strategy. Program design should first focus on desired outcomes and benefits. In Austin's case this included tangible progress to improve air quality, combat climate change, and support affordability goals. Also, the Austin strategy promotes national "fuel independence" and leverages a natural synergy from increased EV adoption, renewable energy penetration, and rapid advancements in autonomous vehicle technologies.

References

1. City of Austin Council Resolution 20140410-024, Net-zero community-wide greenhouse gas emissions by 2050. http://austintexas.gov/sites/default/files/files/Sustainability/Climate/Resolution_No_20140410-024.pdf
2. Nealer R, Reichmuth D, Anair D (2015) Union of concerned scientists research paper, cleaner cars from cradle to grave. http://www.ucsusa.org/clean-vehicles/electric-vehicles/life-cycle-ev-emissions#.WLwvRTsrLmE
3. Kelly Blue Book (2016) Press release, avg price index, transaction prices up 2 percent. http://mediaroom.kbb.com/new-car-transaction-prices-up-2-percent-March-2016
4. United States Department of Energy, Energy Efficiency and Renewable Energy, "Benefits and Considerations of Electricity as a Vehicle Fuel". http://www.afdc.energy.gov/fuels/electricity_benefits.html
5. Shoup, D, University of California Transportation Center (1997) Report "The High-Cost of Free Parking". (PARKING 95% OF TIME). http://escholarship.org/uc/item/4vz087cc#page-1
6. Baumer Z, Popham K, Walls J (2016) Plan submission to Austin City council mobility committee Oct 5, 2016, Fleet electrification study and plan. http://www.austintexas.gov/edims/document.cfm?id=264039
7. United States Environmental Protection Agency (2014) Greenhouse gas emissions fast facts document. https://www.epa.gov/sites/production/files/2016-06/documents/us_ghg_inv_fastfacts2016.pdf
8. US Department of Labor Bureau of Labor Statistics, Consumer Expenditures (2015) https://www.bls.gov/news.release/cesan.nr0.htm
9. Data derived from California ISO Grid Operator (2016) and chart presented on Wikipedia. https://en.wikipedia.org/wiki/Duck_curve
10. United States Energy Information Administration (2015) FAQ: how much petroleum does the United States import and export? http://www.eia.gov/tools/faqs/faq.cfm?id=727

Chapter 8
Smart Transportation Systems

Jesus A. Jimenez

Introduction

The U.S. Department of Transportation defines Intelligent Transportation Systems (ITS) as "a set of tools that facilitates a connected, integrated, and automated transportation system that is information-intensive to better serve the interests of users and be responsive to the needs of travelers and system operators" [1]. The users, either human passengers or physical goods, will be part of one big network made possible by the "Internet of Things" (IoT). Data, such as traffic congestion, road conditions, passenger information, etc. will be collected from IoT-enabled devices. The data will be processed by supercomputers and cloud computing, and then computer algorithms will optimize the transportation resources, user's routes and safety, and the user's experience. As pointed out by Monzon [2], the Smart City infrastructure is the core concept, and the technology is the enabler, but it is the level of connectivity and integration of all systems that will make the system truly smart.

Smart Transportation Components

Before the 1980s, researchers focused on the development of in-vehicle navigation and route guide systems. Many components were developed during these years to enable communications between the vehicle and objects strategically positioned near the vehicle. The first efforts started with the Electronic Route Guidance System

J.A. Jimenez (✉)
Ingram School of Engineering, Texas State University, San Marcos, TX 78666, USA
e-mail: Jesus.Jimenez@txstate.edu

S. McClellan et al. (eds.), *Smart Cities*, DOI 10.1007/978-3-319-59381-4_8

(EGRS) in the 1970s [3, 4]. The system used a unique coding system to identify highway intersections based on geographical representations. Electronic equipment was mounted in vehicles where drivers entered destinations and read signals from the routing instruments installed along the path of the vehicle. Vehicles communicate with these instruments as they move through the route and the instrument transmits routes instructions to the vehicle, displayed in the form of easy to read symbols. A similar system called the Comprehensive Automobile Traffic Control System (CACS), were also tested in urban areas in Japan [5]. CACS used a central traffic control center that communicated traffic data to instruments installed in vehicles. Trials found benefits in traffic reduction since vehicles could avoid congested routes by using this system. The automatic route control system (ARCS) was developed for newspapers' delivery applications [6]. ARCS provided directions as a newspaper delivery vehicle moved in a fixed route. The system gave real-time visual and audio commands to help the drivers find drop off locations as they navigated their routes. The system could also make adjustments to the routes in case significant departures from the fixed route were detected. Automatic vehicle location (AVL) used electromagnetic and ultrasonic sensors and radio signals to track the location of a vehicle. A central computer processed the vehicle location information and displayed positioning data in lighted maps. These systems were popular in the 1970s by law enforcement agencies and police control centers to manage their vehicle fleet and dispatch emergency response services based on the vehicle's proximity to the location where the emergency was taking place [7]. The global positioning systems (GPS) made its debut in the commercial applications in the 1980s. GPS used satellite communication received by a GPS receiver to provide geographical positioning of vehicles [8].

Between the 1980s and 1990s, there was a proliferation of technologies to improve road safety and fuel efficiency. Several traffic management operations were automated as a result of programs developed during these periods. The most relevant programs included the development of automated vehicle surveillance systems, weight-in motion and automated vehicle identification for commercial vehicles, electronic toll roads and dedicated short range communication (DSRC) [8].

During the 2000s, the emphasis is to develop more automated vehicle functions and connectivity between the vehicles and the surrounding infrastructure. As a result of these programs, a plethora of IoT-enabled ITS resources emerged. The infrastructure of IoT comprises mainly a cloud-based host center, gateways, routers, access points, and switches. These devices enable system interoperability, such as the vehicle-to-vehicle (V2V) and vehicle-to-infrastructure (V2I) connectivity. Other physical objects, such as people, machines, and devices, can be connected to the network as well. These objects are capable of exchanging data without any human or computer interaction. The wireless networks enable rapid communications between vehicles and the roadside infrastructure and buildings. The coverage range of the wireless devices may be limited, but a solution to extend the range consists in installing multiple roadside nodes so that the information is passed from one node to another [3].

Other key underlying technologies enablers of ITS are listed in Table 8.1 [3, 9] and described in following subsections.

Table 8.1 Typical components underlying ITS technology

Technology	Description
V2V and V2I	Vehicle-to-vehicle and vehicle-to-infrastructure communications
GPS	Global positioning systems
DSRC	Dedicated short range communications
3G, 4G, 5G, LTE, etc.	Mobile telephony
RCR	Roadside camera recognition
PV/PD	Probe vehicles or devices

V2V and V2I

Direct or indirect communication between two vehicles or between a vehicle and a roadside infrastructure can improve driver's safety and mobility by using the wireless network. According to Maekawa [9], the applications include cooperative driver assistance, decentralized probe vehicles, and user and information communications. For instance, with this technology, vehicles have the capability of sending a warning to other vehicles to avoid accidents while changing road lanes.

GPS

GPS is a core technology behind many in-vehicle navigation and route guidance systems. By using data received from satellites, GPS receivers can accurately track the positioning of the vehicles in a city. The accuracy of the GPS is not the same for pedestrian traffic near highly congested sidewalks or inside buildings where the signals can be shielded or in the way of other multiple transmissions.

Dedicated Short Range Communication

Dedicated short range communications (DSRC) operating in the 5.8 GHz (U.S.) or 5.9 GHz (Japan/Europe) wireless spectrum, DSRC makes two-way wireless connections between vehicles and roadside infrastructure. The applications include V2I integration, V2V communication, adaptive traffic signal timing, electronic toll collection, congestion charging, electronic road pricing, information provision, etc.

3G, 4G, 5G

Vehicles and infrastructure can get connectivity through 3G or 4G mobile networks to expand availability in inaccessible points, such as most rural areas. Cost and network speed may inhibit the use of this type of networks.

RCR

Roadside camera technology is increasingly used for traffic management and accident prevention. For instance, e-toll roads use cameras to read license plates, count number of passengers, count number of axles per commercial trucks, etc. so that the proper usage fees can be collected from the users. Cameras are connected to a central computer where a computer analytics algorithm extracts the information needed from photographs and videos.

PV/PD

Taking advantage of the sensors found in vehicles, a probe information system collects information from these devices and transforms it into usable data. For instance, the speed obtained from a probe vehicle can generate traffic congestion data, or the activation of wipers in the probe cars can generate route weather and rainfall information.

Issues Motivating ITS

Smart transportation seeks to address long-standing issues arising from over-population and population growth, the inefficiencies associated with most conventional transportation resources, and the aging of transportation systems. These issues are discussed in following subsections.

High Traffic Density

According to the United Nation's World Urbanization Prospects report, in 2014, 54% of the world population lives in cities [10]. Private vehicles are integral part of people's daily life and contribute the largest number of vehicle's distribution. According to the report from National Household Travel Survey, the total number of private vehicles reached 113 million in 2009. The average number of vehicle per household increased from 1.16 in 1969 to 1.86 in 2009 [11]. According to statistical reports made available by the U.S. Department of Transportation, Bureau of Transportation Statistics (BTS), the miles of vehicle traveled increased by 40,542 miles, from 228,019 million miles in 2000 to 268,561 million miles in 2016, which represents a 15% increment [12].

Long Transportation Times and High Costs

According to the Smart City Challenge's Lessons Learned Report [13], 10% of the traffic delays in urban centers is caused by outdated traffic signals and 30% of traffic is caused by those cars which drive in the city trying to find parking. According to [11], in the Metropolitan Statistical Area, the average time spent driving a private vehicle per day is 56.09 min. Similarly, the yearly delay per auto commuter in 2014 is about 42 h, whereas the wasted fuel per auto commuter is 29 gallons. That is, the total travel delay nationwide is 6.9 billion hours, whereas the wasted fuel is 3.1 billion gallons nationwide. This leads to a total cost due to congestion equal to US $160 billion.

High Carbon Dioxide Emissions

Transportation activities have significant impact on the environment. Due to the low efficiency of the transportation resources, the transportation industry is the second largest contributor of carbon emissions, which was equivalent to 26% of total greenhouse gas emission in the United States in 2015 [14].

Expanded Supply Chains

According to a recent survey published by the Material Handling Industry of America [15], the supply chain industry has already begun efforts in adopting smart cities solutions, and transportation is the first area in which the supply chain industry plans to take advantage of the smart cities solutions.

According to the America Trucking Association [16], in 2015, there were 3.63 million Class 8 trucks in operation. The statistics also indicates that trucks carried 70.1% of domestic freight tonnage and generated US $726.4 Billion in gross freight revenues. Reports also indicate that trucks in stop-and-go traffic result in wasted fuel and high operating costs, which total $28 million per year [13].

The railway industry in the U.S. seems to be stable. According to reports from Association of American Railroads and U.S. Department of Transportation, Bureau of Transportation Statistics (BTS) [12], the Class I Rail Tonnage is about 1700 million tons.

Challenges

The number of ITS projects is increasing. Looking ahead, there are some challenges that these systems must overcome for increased technology penetration. Some of the challenges of smart transportation are provided below. These are grouped into four categories: Information safety and privacy, Easy access and universality, Time, Jobs, and Funding.

Information Safety and Privacy

Smart transportation paradigm works effectively in part because of the access, analysis, and interpretation of the data transmitted by the ITS elements. Big amounts of data are exchanged between the devices. Some of the data exchanged may contain user's personal information. Such data is to be stored and maintained in cloud environments. One challenge with the data is to protect individual's privacy of end users. For example, companies can develop targeted marketing and their customers will receive personalized ads directly on their mobile devices while driving by near the company's retail stores or business establishments. Solutions to ensure authentication, integrity and confidentiality of cloud-based data and communications are a critical area of research [17].

Coordination, Easy Access and Universality

According to the U.S. Department of Transportation [13], the general public can get free access to transit time data, which are generated and are maintained by 28% of the transit agencies. The challenge is to ensure that all the sectors and systems provide coordinated and user-friendly access to these databases. It is also equally important that these systems are presented in formats that can be understood by end users with different background, education, or skills levels.

Funding

Funding of ITS projects is another realistic challenge that ITS has faced. The Integrated Surface Transportation Efficiency Act (ISTEA) [18] and the Transportation Equity Act [19] were two U.S. federal government initiatives that emerged in the 1990s to support the development of the transportation systems of the future. These programs allocated funding for the development of a prototype of an intelligent vehicle highway system. The Transportation Equity Act followed ISTEA to research, develop, and test ITS. The program also comprises the ITS technology deployment. The Moving Ahead for Progress in the 21st century Act allocates funds to explore electric vehicle charging and electronic toll roads [20]. In the Smart City Challenge, a contest sponsored by the U.S. DOT, funding was provided to several U.S. cities to design smart transportation city solutions for the year 2045 [13]. Among the 78 proposals received, solutions such as dynamic route planning, programmable streets with dynamic markings, on-demand last-mile-service shuttles, and solar-powered electric vehicle charging stations, were presented. For additional details of the proposals received by this program, visit https://www.transportation.gov/smartcity.

Rebuild Road Network

Besides expanding the surface transportation, a more challenging issue is to upgrade the existing infrastructure, which in most cases needs to be redesigned in order to accommodate the requirements of the new environments that ITS developments are originating. Old facilities and buildings may also need to be reconditioned as well to be able to support new technological advancements.

Training the ITS Workforce

The transportation industry has experienced issues to retain their workforce. In the trucking industry, for instance, there is an attrition of drivers, who due to the extensive trips and sacrificed lifestyle prefer to change careers to more local jobs that keep them closer to their homes. The problem with workforce retention is getting more serious with ITS, in part because transportation is expected to become data-driven and highly automated. The challenge is thus to develop a "knowledge-based" workforce of fully skilled workforce in the areas of transportation, supply chain, and IoT who can be capable of performing advanced data collection, analytics and interpretation [21]. The new ITS workforce will be the responsible for the installation, operation and maintenance of the ITS technology in future smart cities.

Major Players in Smart Cities and Transportation

The Smart Cities' market value will be $1.2 billion in 2019, according to Siemens [22]. With 25% of the market share, Industry Automation is the leading segment, whereas smart transportation owns 8%. To evaluate the industry positioning on smart transportation, we analyze several key industry players including 3M, Cisco, IBM, Intel, NEC, Samsung, and Siemens [22–28].

In this section, the analysis focuses on the vision and goals of these companies as related to smart transportation. This information was obtained by conducting a web search using the company plus the keywords "Smart City" and "transportation." The results of this comparison are summarized in Tables 8.2 and 8.3.

Alignment with ITS Goals

The goals explored in this comparison, and summarized in Table 8.2, are as follows:

Table 8.2 Comparison of ITS vision and goals by major industry players

Company	Corporate vision	CON	EFF	MOB	SFT	SUS	USR
Cisco	"Connected transportation: improve safety and the quality of life in cities with increased mobility and efficiency"						
IBM	"New cognitive approaches to long-standing challenges"		X				X
Intel	"Shaping the future of transportation through technology innovation"	X		X	X		
3M	"Building the city of the future"				X		
NEC	"At NEC, we help cities flexibly respond to challenges that arise during each stage of development"					X	X
Samsung	"Smart cities solving parking and driving challenges"	X					
Siemens	"At Siemens we lead the way towards more intelligent and efficient infrastructures"	X	X	X	X	X	X

Table 8.3 Comparison of ITS technologies by major industry players

Company	3M	Cisco	IBM	Intel	NEC	Samsung	Siemens
V2V and V2I		X				X	
GPS		X					
Driver management and scheduling		X					
Fleet maintenance and efficient fuel management		X		X			
Infrastructure connectivity		X					X
IoT gateways		X		X			
IoT platform and cloud computing		X	X	X			
IoT-based supply chain management			X	X			
Real-time vehicle monitoring	X	X					
Roadside cameras and toll roads	X	X			X		

- Connectivity of vehicles and infrastructure (Code: "CON")
- Efficiency and productivity of users (Code: "EFF")
- Mobility (Code: "MOB")
- Safety (Code: "SFT")
- Sustainability (Code: "SUS")
- User experience (Code: "USR").

From the analysis and summary in Table 8.2, it is clear that only one company articulates vision and goals with a focus on the key areas of ITS which have been described previously (Siemens). Other competitors articulate vision/goals which align with a subset of ITS key areas, such as "Connectivity of vehicles and infrastructure", "Mobility", and "Safety" (CON + MOB + SFT, Intel), "Sustainability" and "User Experience" (SUS + USR, NEC), or "Efficiency and productivity of users" and "User Experience (EFF + USR, IBM).

Alignment with ITS Technologies

The technologies related to ITS which are explored in this comparison are summarized in Table 8.3. The technologies include

- V2V & V2I
- GPS

- Driver Management and Scheduling
- Fleet Maintenance and Efficient Fuel Management
- Infrastructure Connectivity
- IoT Gateways
- IoT Platform and Clout Computing
- IoT-based Supply Chain Management
- Real-time Vehicle Monitoring
- Roadside Cameras and Toll Roads.

From the analysis and summary in Table 8.3, it is clear that only one company articulates technologies with a focus on the key areas of ITS which have been described previously (Cisco). Other competitors articulates technologies which align with a subset of ITS key areas, such as "Real-time Vehicle Monitoring" and "Roadside Cameras and Toll Roads" (3M), "IoT Platform and Clout Computing" and "IoT-based Supply Chain Management" (IBM), or "Fleet Maintenance and Efficient Fuel Management," "IoT Gateways," "IoT Platform and Clout Computing," and "IoT-based Supply Chain Management" (Intel).

Conclusion

This chapter has discussed Intelligent Transportation Systems (ITS), along with many of the key considerations, policies, issues, and technologies which shape this area of Smart City development. An analysis of key players in the industry is presented which compares the visions and goals of the companies with key aspects of ITS as well as the key technologies of ITS. As conventional surface transportation resources continue to be heavily used and continue to age, transportation systems will continue to adapt, changing the way humans and products move around the new smart cities of the future. Autonomous vehicles will communicate with other vehicles and with city infrastructure, profoundly transforming the transportation industry and its constituencies. Smart cities which employ and effectively manage these technologies will improve mobility while reducing pollution, and in so doing will improve quality of life and economic productivity for their citizens.

Acknowledgement The author would like to thank Mr. Yue Li for his valuable contributions.

References

1. U.S. Department of Transportation (2014) USDOT's intelligent transportation systems (ITS) ITS Strategic Plan 2015–2019. https://www.its.dot.gov/strategicplan.pdf. Accessed 29 Mar 2017
2. Monzon A (2015) Smart cities concept and challenges: bases for the assessment of smart city projects. In: International conference on smart cities and green ICT systems (SMARTGREENS) May 2015

3. An S-H, Lee B-H, Shin D-R (2011) A survey on intelligent transportation systems. In: Third international conference on computational intelligence, communication systems and networks, July 2011
4. Rosen DA, Mammano FJ, Favout R (1970) An electronic route-guidance system for highway vehicles. IEEE Trans Veh Technol 19:143–152
5. Totani S (1980) Development and current status of CACS (comprehensive automobile traffic control system). In: 30th IEEE vehicular technology conference, Sept 1980
6. French RL, Lang GM (1973) Automatic route control system. IEEE Trans Veh Technol 22:36–40
7. Riter S, McCoy J (1977) Automatic vehicle location—an overview. IEEE Trans Veh Technol 26:7–10
8. U.S. Department of Transportation Intelligent Vehicle Systems Joint Program Office (2016) History of intelligent transportation systems. https://www.its.dot.gov/history/index.html. Accessed 7 Apr 2017
9. Maekawa M (2004) ITS (Intelligent Transportation Systems) Solutions. NEC J Adv Technol 1:194–199
10. United Nations, Department of Economic and Social Affairs, Population Division (2014) World urbanization prospects: the 2014 revision, highlights (ST/ESA/SER.A/352)
11. U.S. Department of Transportation, Federal Highway Administration (2011) Summary of travel trends: 2009 national household travel survey. http://nhts.ornl.gov/2009/pub/stt.pdf. Accessed 29 Mar 2017
12. U.S. Department of Transportation Bureau of Transportation Statistics (2017) National transportation statistics. https://www.rita.dot.gov/bts/data_and_statistics/index.html. Accessed 29 Mar 2017
13. U.S. Department of Transportation (2015) Smart city challenge: lessons for building cities of the future. https://www.transportation.gov/policy-initiatives/smartcity/smart-city-challenge-lessons-building-cities-future. Accessed 6 Apr 2017
14. US Environmental Protection Agency (2017) Greenhouse Gas Emissions.https://www.epa.gov/ghgemissions/sources-greenhouse-gas-emissions#transportation. Accessed 29 Mar 2017
15. Deloitte Consulting (2017) Material handling industry annual industry report. http://MHI.org/publications/report. Accessed 5 Apr 2017
16. American Trucking Association (2016) ATA American trucking trends 2016.http://www.trucking.org. Accessed 29 Mar 2017
17. Zissis D, Lekkas D (2012) Addressing cloud computing security issues. Future Gener Comput Syst 28:583–592
18. U.S. Department of Transportation (1991) Intermodal surface transportation efficiency act of 1991—Summary. https://ntl.bts.gov/DOCS/ste.html. Accessed 29 Mar 2017
19. U.S. Congress (1997) Transportation equity act for the 21st century. https://www.congress.gov. Accessed 29 Mar 2017
20. U.S. Department of Transportation (2012) Moving ahead for progress in the 21st century act. https://www.fhwa.dot.gov/map21/docs/map21_summary_hgwy_provisions.pdf. Accessed 29 Mar 2017
21. Miller C (2017) Smart cities and their potential impact on supply chains. In: MHI solutions, April–June 2017
22. Siemens (2017) Smart cities: city intelligence platform. http://www.siemens.com. Accessed 8 Apr 2017
23. Cisco (2016) Connected transportation. http://www.cisco.com. Accessed 7 Apr 2017
24. IBM (2017) Smart cities. http://www.ibm.com. Accessed 8 Apr 2017
25. Intel (2016) Connected transportation and logistics. http://www.intel.com. Accessed 7 Apr 2017
26. NEC (2017) Smart cities. http://www.nec.com. Accessed 8 Apr 2017
27. Samsung (2017) Insights. https://insights.samsung.com. Accessed 8 Apr 2017
28. Hoffman K (2015) Building the city of the future. http://www.3m.com. Accessed 7 Apr 2017

Chapter 9
Reconfigurable Computing for Smart Vehicles

Vikas Chaudhary

Introduction

The introduction of electronics for vehicular computation began in the early 1980s with lightweight micro-controllers and point-to-point wired connections for tasks like fuel injection control, antilock braking systems (ABS), and other control functions. The addition of more functions resulted in a large number of point-to-point links, which became complicated and difficult to manage. Moreover, the amount of wiring began to contribute to the weight of the car. These issues have led to the development of network-based solutions as well as embedded architectures to improve cost, efficiency and performance. Recently, large sets of advanced functionality have appeared in Advanced Driver Assistance Systems (ADAS). These systems often include assistance based on camera and radar inputs, such as brake and park assistance, lane departure detection, night vision assistance with pedestrian recognition, or suspensions which proactively scan the road surface. In the future, vehicles will become entities of large intelligent transportation systems, involving cooperation mechanisms through vehicle-to-vehicle (V2V) and vehicle-to-infrastructure (V2I) communications [1].

It is estimated that modern, well-equipped automobiles use more than 50 microcontroller units (MCUs) [2]. Some high-end vehicles contain as many as 100 ECUs (Electronic Control Units) [3]. These MCUs perform many functions related to safety, energy savings, passenger comfort, convenience, entertainment, and many others. These MCUs communicate with each other through in-vehicle network systems. As a result, in-vehicle networking has evolved at a fast pace in the automotive industry, resulting in the elimination of wiring harnesses once used in control units. Current in-vehicle networking includes onboard communications systems, control systems such as lane departure warning, and road condition

V. Chaudhary (✉)
ON Semiconductors, Austin, TX 78746, USA
e-mail: vikascha@gmail.com

S. McClellan et al. (eds.), *Smart Cities*, DOI 10.1007/978-3-319-59381-4_9

advisories [1]. Currently, innovations within the automotive domain are driven by embedded systems and software solutions. Many of these innovations like antilock braking systems, electronic stability control, or emergency brake assistance significantly reduce vehicle accidents and thus increase safety. On the other hand, embedded systems enhance the driving comfort with driver assistance functions like adaptive cruise control and auto climate control. Furthermore, infotainment systems and telematics enhance the user's driving experience [2].

As the technology is advancing rapidly towards autonomous driving, an exponential growth in electronics usage will ensue. Analysts predict that the market for semiconductors in automotive applications will increase at a compound annual growth rate of 7% in the next five years, and with ADAS systems this growth rate might exceed 20% [4]. The cost of electronics can comprise as much as 50% of the total vehicle cost, as shown in Fig. 9.1 [5]. New cars will be heavily equipped with a large number of sensors, antennas and image cameras. Fast computing power will be required to process such a large volume of data. The best way to achieve fast data processing is to offload the majority of this work to hardware accelerators. The standard MCUs used in cars today are based on hardware platforms provided by chip manufacturers. These basic systems may lack the flexibility to address specific needs of the customer. If an auto manufacturer wants to add a new hardware feature based on some specific application, it is generally difficult and time-consuming to respond to such a request quickly. However, these application-specific hardware requirements may be fulfilled through the use of Field-Programmable Gate Arrays (FPGA) where massive customization and parallelism is possible [6].

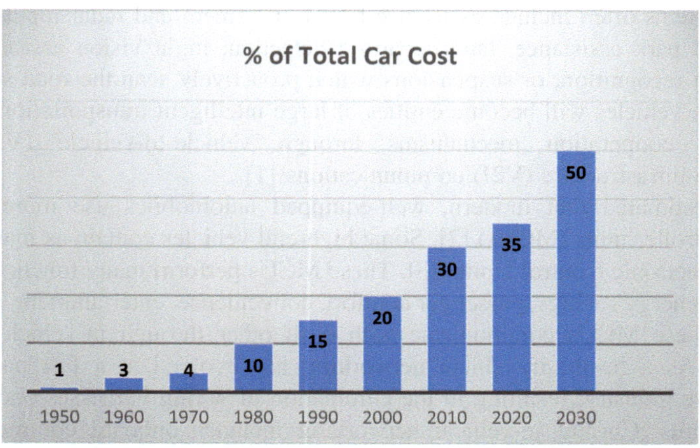

Fig. 9.1 Electronics cost projections

Automotive Communication Systems

In modern vehicles, various devices/systems communicate using multiple networking protocols. Vehicle communication can be broadly classified into three major categories, which are discussed in more detail in following sections: In-Vehicle, Vehicle-to-Vehicle (V2V), and Vehicle-to-Infrastructure (V2I). All these communication channels help in gathering information about conditions surrounding the vehicle, such as road, weather, object/pedestrian detection, and traffic conditions. Adequately processing this information enables vehicle, passenger, and environmental safety. The processed information can be used to alert or warn the driver in semi-autonomous (command control) driving, or be used in making decisions for fully autonomous driving.

In-Vehicle Communication

Early vehicular electronics consisted of simple devices like 8-bit micro-controllers with simple I/O support to connect to the sensors. With the introduction of more demanding applications, more powerful 16- and 32-bit processors and domain-specific controllers came into use to provide improved computational capabilities. The primary advantage of general-purpose processors is software portability and independence from underlying hardware due to abstractions supported by high-level languages. This trend continued in the standardization of the requirements and capabilities of underlying operating systems which enable application developers to design their product independent of the hardware target. AUTOSAR (AUTomotive Open System ARchitecture), which is based on OSEK, was designed to provide a standard software architecture for the various electronic control units (ECUs) throughout a vehicle. AUTOSAR concepts are widely adhered to in the automotive industry, and determine a standardized platform for automotive applications [7].

Early automotive systems used simple switches and actuators, and their functionality was implemented using point-to-point wiring. As new and more advanced features were introduced, point-to-point connections became infeasible due to the complexity of the wiring harness. To mitigate problems due to point-to-point wiring, Robert Bosch introduced the Controller Area Network (CAN) in 1986 [8], which gained widespread acceptance in the automotive industry and later became the most widely used networking backbone for in-vehicle systems. A CAN network connects a vehicle's electronic equipment similar the familiar LAN connections of personal computers. These networks facilitate the sharing of information and resources among the distributed applications [9]. A typical vehicle can contain two or three separate CANs operating at different transmission rates. A low-speed CAN typically runs at less than 125 Kbps and manages less critical features such as seat adjustments, window movement controls, etc. A higher-speed CAN runs more real-time-critical functions such as engine management, antilock brakes and cruise control [9].

Another in-vehicle network called LIN (local interconnect network) was developed in the late 1990s to reduce CAN cost for less critical functions. LIN applications typically include communications between intelligent sensors and actuators, such as window controls, door locks, rain sensors, windshield wiper controls, and climate control. The LIN bus is a low-cost, serial-bus communication protocol that has single master and one or more slaves. All messages are initiated by the master with only one slave responding to each message at a time. In such an implementation, there is no need for collision detection and arbitration. The transmission packet size varies from 1 to 8 bytes and takes place through a Universal Asynchronnous Receiver/Transmitter/Serial Communications Interface (UART/SCI). Data integrity is maintained using checksums. Data rates in the LIN bus range from 1 to 20 kb/s, which is suitable for the intended applications and minimizes electromagnetic interference. The LIN bus is always in one of two states: active or sleep. When it is active, all nodes on the bus are awake and listening for relevant bus commands.

As there were further advances in automotive electronics, automobile manufacturers and leading suppliers created new automotive network scheme called FlexRay which is better suited for advanced applications in automation [8]. FlexRay delivers faster speed with maximum data rate of 10 Mbps, which is faster than CAN and can meet the performance requirements for X-by-Wire applications (i.e., drive-by-wire, steer-by-wire, brake-by-wire, etc.) [8].

FlexRay is a differential bus running at speeds up to 10 Mb/s, which is significantly faster than LIN (20 kb/s) or CAN (1 Mb/s). FlexRay uses a dual-channel architecture that has two major benefits. First, the two channels can be configured to provide redundant communication in safety-critical applications to ensure the message is delivered. Second, the two channels can be configured to send unique information on each link at 10 Mb/s, giving an overall bus transfer rate of 20 Mb/s in less safety-critical applications [10].

A typical vehicle communication system interface is shown in Fig. 9.2.

Vehicle-to-Vehicle (V2V) Communication

V2V is a crash avoidance technology which relies on information communicated between nearby vehicles to warn drivers about dangerous situations that can lead to an accident. The National Highway Traffic Safety Administration (NHTSA) is a US government agency which helps to reduce deaths, injuries, and economic losses resulting from motor vehicle crashes by setting and enforcing safety performance standards for motor vehicles and motor vehicle equipment [11]. The NHTSA's emphasis on adoption of crash avoidance technologies, like electronic stability control, has helped vehicles react to crash-imminent situations, but has not yet been able to help the driver react ahead of time. To address driver reaction issues, some of

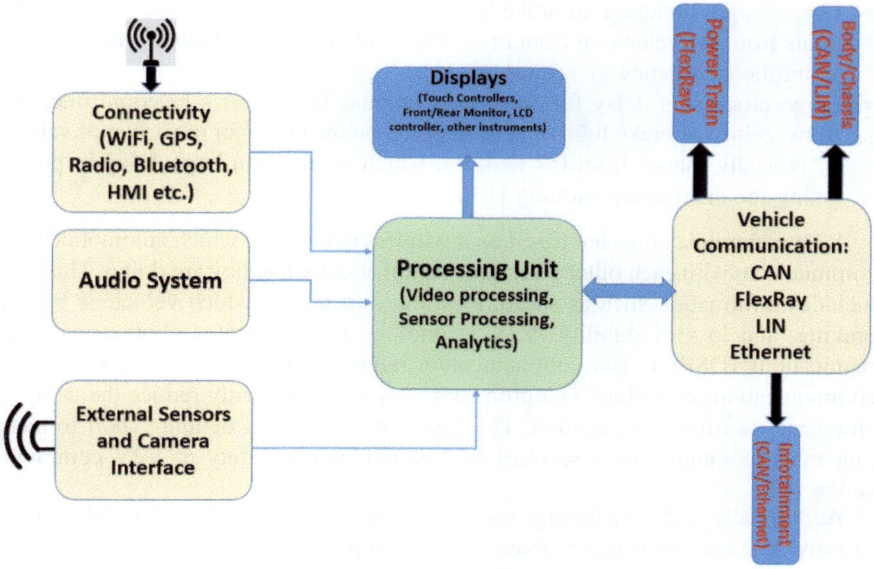

Fig. 9.2 Typical in-vehicle communication system

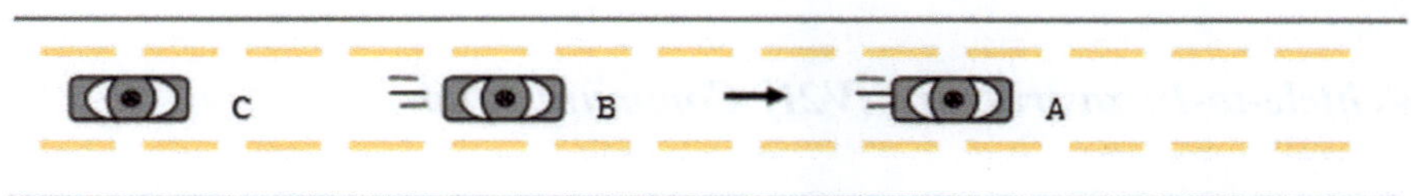

Fig. 9.3 Vehicle-to-vehicle communication

the most advanced crash avoidance technologies available today include a host of onboard sensors, cameras, and radar applications as shown in Fig. 9.2. These technologies may warn of impending danger so that the driver can take corrective action. Some technologies may even be able to intervene on the driver's behalf [11].

V2V communications represent an additional step in helping to warn drivers about impending danger [11]. V2V communication will be an additional step toward intelligent transportation systems to enhance road safety. Over the past decade, V2V communications have attracted a lot of attention and currently various applications exist, including cooperative forward collision warning, traffic light optimal speed advisory, and remote wireless diagnosis [12].

Human drivers suffer from perception limitations in roadway emergency events, resulting in large delays in propagating emergency warnings. Such a scenario is described by Yang [13]. Yang describes a scenario where there are three vehicles A, B and C traveling in the same lane, as shown in Fig. 9.3. If A suddenly brakes, both vehicles B and C are endangered, and being further away from A does not make vehicle C any safer than B due to the following:

- Line-of-sight limitation of brake lights: Typically a driver can only see the brake lights from the vehicle in front of it. Thus, the driver in vehicle C will not know about the emergency of vehicle A [13].
- Large processing delay for emergency events: The driver's reaction time, i.e., from seeing the brake light of A to stepping on the brake for the driver of vehicle B, typically ranges from 0.7 to 1.5 s, which results in a large delay in propagating the emergency warning [13].

V2V communications are based on a wireless network in which automobiles can communicate with each other with information about what they are doing. This data includes information such as speed, location, direction in which vehicle is bound, braking, and loss of stability. V2V technology uses dedicated short-range communications (DSRC). The communication range of DSRC is up to 300 m. This communication technology is a promising way to significantly reduce the delay in emergency warning propagation. The DSRC consortium is defining short to medium range communication services that support public safety in V2V communication [11].

Additionally, V2V technology can be combined with existing radar and cameras to provide more information about the surrounding traffic conditions. This combined approach with enhanced information can improve decision making, especially for driverless vehicles.

Vehicle-to-Infrastructure (V2I) Communication

V2I communication is an extension of V2V communication to augment safe driving and improve collision avoidance. V2I communications involve the wireless exchange of critical safety and operational data between vehicles and roadside infrastructure. V2I communications are primarily intended to avoid motor vehicle crashes [11]. V2I communications apply to all vehicle types and all roads, and transform infrastructure equipment into "smart infrastructure" through the communication between vehicles and infrastructure elements to perform calculations that recognize high-risk situations in advance, resulting in alerts/warnings [11]. A typical V2I communication network is shown in Fig. 9.4.

V2I applications complement the V2V safety applications by addressing crash scenarios that V2V applications cannot address, and by more efficiently addressing some crash scenarios when there are low levels of penetration of DSRC-equipped vehicles. The following is a list of contemplated V2I safety applications [11]:

- **Red Light Violation Warning**: This technology will provide in-vehicle alerts to drivers about potential violations of upcoming red lights, based on vehicle speeds and distances to intersections.

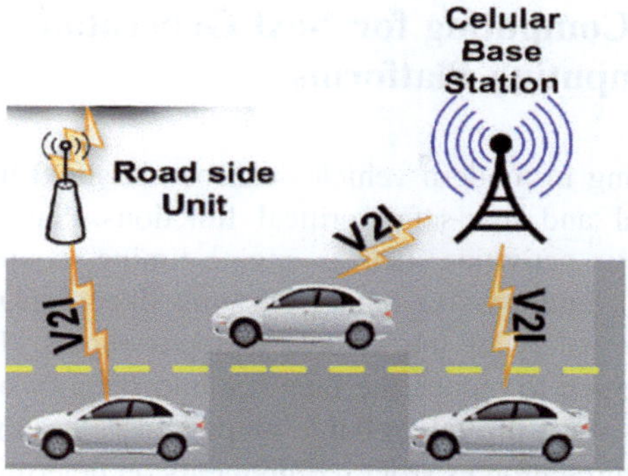

Fig. 9.4 Vehicle-to-infrastructure communication (V2I)

- **Curve Speed Warning**: If a driver's current speed is unsafe for traveling through an upcoming road curve, this technology will alert the motorist to slow down.
- **Stop Sign Gap Assist**: This technology will assist drivers at STOP-sign-controlled intersections via vehicle gap detections, alerting motorists when it is unsafe to enter intersections.
- **Reduced Speed Zone Warning**: This technology will assist drivers in work zones, by issuing alerts to drivers to reduce speed, change lanes, and/or prepare to stop.
- **Spot Weather Information Warning**: This technology will provide in-vehicle alerts or warning to drivers about real-time weather events and locations, based upon information from Roadside Equipment connections with Transportation Management Center and other weather data collection sites/services.
- **Stop Sign Violation Warning**: Based on vehicle speeds and distances to intersections, this technology will provide in-vehicle alerts to drivers about potential violations of upcoming stop signs.
- **Railroad Crossing Violation Warning**: This technology will assist drivers at controlled railroad crossings via connections with existing train detection equipment, alerting motorists when it is unsafe to cross the railroad tracks.
- **Oversize Vehicle Warning**: Drivers of oversized vehicles will receive an in-vehicle alert to take an alternate route or a warning to stop, based upon information from connections to infrastructure at bridges/tunnels.

Vendors like Qualcomm have come up with automotive platforms to support V2V and V2I communications. Qualcomm's Snapdragon 820A platform claims to integrate V2X (Vehicle-to-Vehicle, Vehicle-to-Infrastructure, Vehicle-to-Pedestrian and Vehicle-to-Cloud) through combination of Wi-Fi, DSRC and LTE [14].

Reconfigurable Computing for Next Generation Automotive Computing Platforms

Automotive computing in modern vehicles can be classified into two major categories: safety-critical and non-safety-critical functions. Safety-critical functions control and coordinate activities such as engine timing, controllability (steering, brakes, acceleration), and passive safety systems like airbags. These functions directly relate to the safety of the vehicle and passengers. These safety-critical mechanisms have helped to reduce the fatalities and severity of vehicle accidents. Hence, such systems must provide reliable outputs at deterministic time intervals. This time-critical nature enforces strict requirements on determinism and reliability of the output generated by embedded control units. On the other hand, non-safety-critical systems are applications that augment the user experience in the vehicle. These functions include comfort-related activities such as seat settings, multimedia, climate control, or assistance activities such as cruise control and remote diagnostics [15].

As technology progresses rapidly towards semi-autonomous (command control) and eventually to fully autonomous vehicles, the computation requirements will grow exponentially. Advanced driver assistance systems (ADAS), used in modern cars, keep an eye on the road and the surroundings with the help of multiple sensors and cameras to keep the driver aware and alert when conditions are not safe for driving. Current ADAS solutions only provide a method of warning the driver of potential threats such as lane departure warning, parking assist, blind spot detection, in-lane vehicle distance warning, and emergency braking. The new features such as active steering control and self-parking can extend the vehicles to semi-autonomous driving. Eventually these advances are going to enable fully autonomous driving where there will be no driver intervention. The timeline of such progression and sensor-usage requirement projections are shown in Fig. 9.5 [8]. Fully autonomous vehicles will have to process information based on in-vehicle sensors and images as well as information from other vehicles (V2V) and roadside infrastructure (V2I), which includes GPS and LTE data.

Auto manufacturers like Tesla with their Model S and Model X are testing their fully self-driving car features [16]. Tesla's model car is equipped with eight surround cameras to provide 360° visibility around the car at up to 250 m. This model also has twelve ultrasonic sensors to complement this vision, allowing the detection

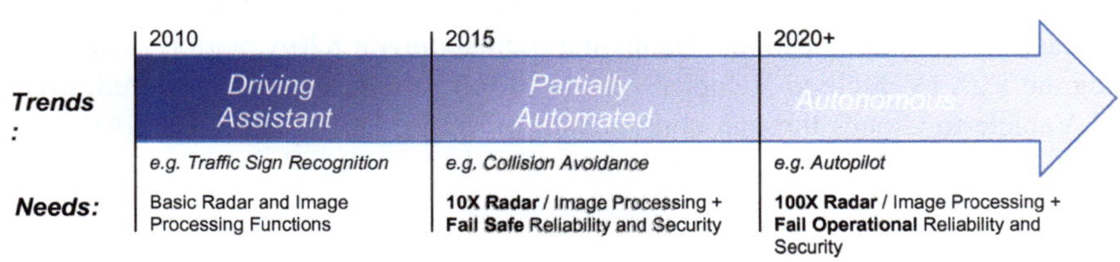

Fig. 9.5 Sensor usage projection in automotive applications

of both hard and soft objects. A forward-facing radar provides additional data about the surrounding on a redundant wavelength, capable of seeing through heavy rain, fog, dust [16]. The Tesla autopilot system is based on Nvidia Graphics Processing Unit (GPU) and CUDA parallel computing platform [17]. Similarly, Google has displayed a self-driving car prototype based on LiDAR (Light Detection and Ranging) as its major sensor. LiDAR is a surveying technology that measures distance by illuminating a target with a laser light. This system weighs roughly 80 kg, is very expensive, and must be mounted on top of the vehicle to achieve a clear line of sight [18]. Current implementations have ranges up to 150–200 m. LiDAR works well in lighted conditions, but starts failing with increases in snow, fog, rain, and dust particles in the air due to its use of light spectrum wavelengths. LiDAR cannot detect color or contrast, and cannot provide optical character recognition capabilities [18].

As mentioned in the examples related to Tesla and Google, ADAS requires massive computing power to process the data generated by a multitude of sensors such as image sensors, radar, LiDAR etc. These systems require real-time data processing to assist the driver in semi-autonomous operations. The computational requirement for such systems will exponentially increase when the industry moves towards fully autonomous driving where there will be a 100× increase in the number of sensors as shown in Fig. 9.5 [8]. This type of data processing coming from various sources uses heterogeneous computing elements such as CPUs, GPUs, and image cognitive processors. Additionally, future systems will incorporate V2V and V2I communications.

The future automotive system will be characterized by a high volume of data on which the vehicle system must perform complex computations in a time-critical manner. Current platforms such as NXP's S32R27, which is based on a 32-bit Power Architecture, is designed to address advanced RADAR signal processing capabilities merged with microcontroller capabilities for generic software tasks and car bus interfacing [8]. However, the performance required to process high volumes of raw data coming from various interfaces is a difficult task for a general-purpose microprocessor. Generally, microprocessors are not optimized for a specific task or data type. For example, GPUs can perform image processing and scientific computing faster than a general-purpose microprocessor due to their application-specific architecture. This is an area where FPGAs present an ideal implementation platform. The computational power of custom hardware on FPGAs enables applications that would otherwise be infeasible on software-based systems handled by a general-purpose microprocessor. Further, FPGA implementations are capable of time-critical processing tasks. Designers can split an FPGA into multiple slices to perform individual tasks on each slice, maintaining the predictability of each while ensuring complete isolation between them.

Until now, the use of FPGAs in vehicle applications has been limited to non-critical driver assistance and multimedia functions [1]. There are two primary concerns regarding the use of FPGAs in critical automotive systems: (1) the protection of valid FPGA configurations used for initialization, and (2) the prevention of SRAM corruption during device operation. Unless these concerns are addressed,

FPGAs cannot be part of an ultra-reliable automotive system design [19]. However, there are multiple advantages of using an FPGA [20]:

- Customized sensor interfaces and various IP (Intellectual Property) cores to enable connectivity to any automotive network standard.
- Ability to implement customized algorithms in hardware and software for video and image processing.
- Ability to design the exact type and number of communications interfaces needed for ADAS applications.

Currently, the driver assistance solutions are implemented using multiple devices: image capture and pixel processing are performed in hardware by ASIC (Application-Specific Integrated Circuit) and serial object processing is generally performed in software on a digital signal processor (DSP). A traditional microprocessor or GPU can handle frame-level processing and vehicle communication. This multichip implementation can potentially create latency due to off-chip interaction. FPGA vendors like Xilinx have developed automotive grade FPGAs such as the Zynq™-7000 programmable SoCs that are suited for the high computation requirements of ADAS. The combination of programmability of hardware and software allows ADAS imaging flow from outside image sensing [6]. An FPGA like the Xilinx Zynq-7000 can handle all of the above-mentioned functions on a single SOC (System-on-Chip) as shown in Fig. 9.6 [6].

ADAS design must also be reliable and fail-safe. System failure in an autonomous vehicle can occur due to many factors, including software glitches, hardware or sensor failures, incorrect object detection, and communication failure. For example, human intervention helped Google's self-driving car to avoid accidents on multiple occasions [21]. Extreme weather conditions, a broad range of temperatures, and the presence of thunderstorms and lightning can result in hardware/communication failure causing a glitch in the system. In addition to the climate, other environmental conditions that these devices must endure include heat, vibrations and noisy communication channels. Hardware also suffers from reliability issues due to factors such as cross-talk, power supply variation, electromagnetic interference, soft errors, and thermal noise. These noise elements

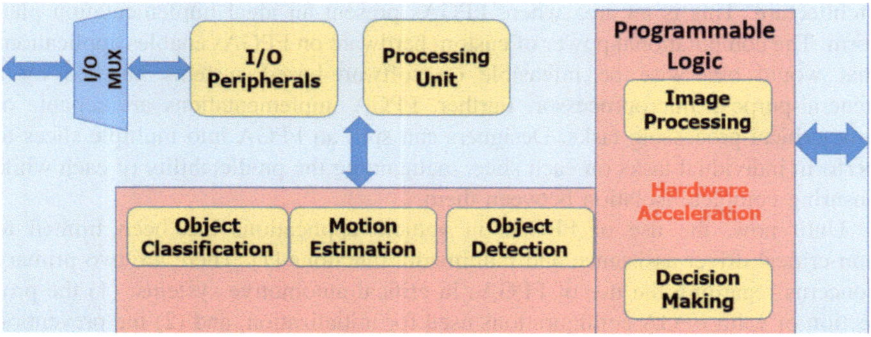

Fig. 9.6 Programmable SOC showing multiple processing units on a single chip

become a dominating factor in a scaled process. Memory elements such as SRAM, latches and flip-flops are especially susceptible to failures. These components are extremely important as memories which store machine state. There are multiple ways in which memories can fail, including program disturbance from tunneling and hot-electron injection, noise effects, erratic tunneling, data retention and read disturbance. Currently, some ECC (Error Correction Code) mechanisms are in place to correct single error in SRAMs. However, this might not be sufficient for the high reliability that is required for automotive applications.

There are many ways to make a system fault-tolerant. From the software perspective, techniques such as N-version programming is common, where many programs performing the same function are executed in parallel and their results matched and voted. Another approach to fault-tolerant system design is dynamic recovery mechanisms. In dynamic recovery, when a fault is detected, the system rolls back to the last known good point and the system restores and starts the computation again. Similarly, from the hardware perspective, fault tolerance is achieved when designers implement circuits so that a chip is not susceptible to PVT variation and can endure high-temperature conditions. A hardware fault-tolerant system is partitioned into fault-containment regions and each module is backed up with redundancy. With this approach, a spare module can assume the function of a failed module. Another method used in hardware is fault masking. A fault masking system is comprised of structural redundancy that masks the faults within a set of redundant modules. Triple modular redundancy is the most commonly used form of fault masking in which the circuitry is triplicated and a majority voting mechanism is used to resolve the results. The voting circuitry can also be triplicated to correct the individual votes.

All the techniques mentioned above require additional hardware and software resources. Moreover, the system should be responsive enough to perform complex computation and take decisions/actions in a time-critical manner. As most of the techniques require some kind of redundancy and additional checking, an FPGA-based solution can be configured to replicate the circuit on a single SOC. Since FPGAs are multi-threaded, all the replicated circuits can be executed in parallel and yield the result at hardware speed. The software-based systems can be slow and might not meet the computation requirement essential in a smart and accurate ADAS system.

FPGAs can play a vital role in such applications as the auto manufacturers can implement more error-correcting features including redundancy and sophisticated error correction algorithms (like BCH or convolutional codes) without going to the platform vendor and thus reducing time to market. Since these features will be under the manufacturer's control with an FPGA-based system, the manufacturers will be able to distinguish themselves from competitors. Moreover, for V2X communication to happen over relatively long distances and under noisy conditions (extreme weather conditions or other interference), reliability via error detection and correction becomes imperative for the fully autonomous vehicle.

Conclusion

Modern vehicles are equipped with intelligent systems to enhance safety. These systems, commonly known as Advanced Driver Assistance Systems (ADAS), gather information from multiple interfaces such as sensors, RADAR, cameras, wireless communication using LTE, and GPS. This information is processed and used in decision making which can warn the driver (semi-autonomous case) or intervene on the driver's behalf. As the technology progresses towards fully autonomous vehicles where there will be no human intervention, exponential growth in the number of sensors, RADAR/LiDAR and cameras integrated to the system is anticipated. This will result in an increased demand on the computing platform, which must support time-critical decisions and must perform appropriate intervention. FPGA-based platforms present an ideal solution for such applications. FPGA platforms can be configured to perform multiple tasks such as image processing, real-time analytics, and sensor data processing on a single SOC. Manufacturers can also implement error detection and correction mechanisms on FPGA to enhance reliability, which is a critical feature for health and safety functions. Since FPGAs are programmable, there is no dependency for the auto manufacturers to rely on platform vendors and hence improve their time to market. Since programming the FPGA is in the manufacturer's control, it enables vendors to distinguish themselves from the competition and to reduce their dependency on specific platforms or subsystems.

References

1. Shreejith S (2016) Enhancing automotive embedded systems with FPGAs. Ph.D. thesis, Nanyang Technological University, Singapore
2. Fleming B (2011) Microcontroller units in automobiles. IEEE Veh Technol Mag 6(3):4–8
3. Navet N, Song Y, Simonot-Lion F, Wilwert C (2005) Trends in automotive communication systems. Proc IEEE 93(6):1204–1223
4. Yoshida J (2013) News & analysis: auto market challenges chipmakers' stamina. http://www.eetimes.com/document.asp?doc_id=1319666. Accessed 25 Mar 2017
5. Chitkara R, Ballhaus W (2013) Spotlight on automotive: PwC semiconductor report. https://www.pwc.com/gx/en/technology/publications/assets/pwc-semiconductor-survey-interactive.pdf. Accessed 15 Mar 2017
6. Xilinx Corporation (2017) www.xilinx.com. Accessed 10 Mar 2017
7. AUTOSAR: The Automotive System Development Partnership (2017) www.autosar.org. Accessed 27 Mar 2017
8. NXP Semiconductors (2017) www.nxp.com. Accessed 12 Mar 2017
9. Leen G, Heffernan D (2002) Expanding automotive electronic systems. Computer 35(1):88–93
10. Davis G (2012) Debugging automotive serial buses: CAN, LIN and FlexRay exposed. http://www.edn.com/design/test-and-measurement/4391269/Debugging-automotive-serial-buses–CAN–LIN-and-FlexRay-exposed. Accessed 3 Mar 2017
11. National Highway Transportation Safety Administration (2017) https://www.nhtsa.gov. Accessed 22 Mar 2017

12. Zhao Z, Cheng X (2014) IEEE 802.11 p for Vehicle-to-Vehicle (V2V) Communications. https://pdfs.semanticscholar.org/c670/9a0df8f6bb02e6fbaff46d2dc179823a8225.pdf
13. Yang X et al (2004) A vehicle-to-vehicle communication protocol for cooperative collision warning. In: The first annual international conference on mobile and ubiquitous systems: networking and services, 2004. MOBIQUITOUS 2004. IEEE
14. Qualcomm Inc. (2017) Snapdragon 820 automotive processor. https://www.qualcomm.com/products/snapdragon/processors/820-automotive. Accessed 15 Mar 2017
15. Shreejith S, Fahmy S, Lukasiewycz M (2013) Reconfigurable computing in next generation automotive networks. IEEE Embed Syst Lett 5:12–15
16. Tesla Motor Company (2017) All Tesla cars being produced now have full self-driving hardware. https://www.tesla.com/blog/all-tesla-cars-being-produced-now-have-full-self-driving-hardware. Accessed 2 Mar 2017
17. Lambert F (2016) Tesla's Autopilot chip supplier NVIDIA on new self-driving system: 'it's basically 5 yrs ahead and coming in 2017'. https://electrek.co/2016/11/11/tesla-autopilot-chip-supplier-self-driving-hardware-its-five-years-ahead. Accessed 5 Mar 2017
18. Barnard M (2016) Tesla & Google disagree about LIDAR—which is right? https://cleantechnica.com/2016/07/29/tesla-google-disagree-lidar-right. Accessed 1 Mar 2017
19. Howell K (2009) Design how-to: tips and tricks: using FPGAs in reliable automotive system design. http://www.eetimes.com/document.asp?doc_id=1272913. Accessed 1 Apr 2017
20. Altera Corporation (2017) Advanced driver assist systems. https://www.altera.com/solutions/industry/automotive/applications/drive-assistance/driver-assist.html. Accessed 24 Mar 2017
21. Korosec K (2016) Humans saved Google's self-driving cars from 13 accidents. http://fortune.com/2016/01/13/google-self-driving-car-accidents. Accessed 15 Mar 2017

Part IV
Infrastructure & Environment

Chapter 10
Smart Buildings and Grid Distribution Management

Phil Powell

Introduction and Overview

This chapter is a case study of successfully building a system that uses the smart building techniques such as the internet of things (IoT) to reduce costs and increase the value of the electrical delivery system supplying power to multiple commercial building processes. I will state at the outset that smart building techniques have been used successfully in most industrial processes by the local plant engineering groups and this chapter is not covering the industrial plant. This chapter focuses on the medium to large commercial buildings that make up everything from strip malls and school campuses to business parks and high-rise offices that are not associated with manufacturing process equipment.

Figure 10.1 shows the main components for delivering electrical energy for processes required by buildings and campuses. Electricity delivery is an engineer-and-forget technology in most of these buildings today. The demand and energy capacity for the building electrical system is forecasted and local codes are used to size the system in a way that will hopefully last the lifetime of the building facility. National standards are used to determine the level of voltage needed for equipment and the design is conservatively produced that will meet these requirements when the electricity supplier, utility or municipal or coop, delivers the national standard voltage range to the building electrical entrance. Once this is done the equipment can be plugged in and the electricity forgotten about as long as the owner is willing to pay the demand and energy costs at the end of the month.

Electrical energy is metered by the supplier with a device that usually tracks hourly demand for the entire facility. The power flows through panel boards with electrical circuit breakers that only interrupt the power flow for failures in the electrical path which is very rare. The entire system only requires intervention when

P. Powell (✉)
Dominion/DVI Grid Solutions, Richmond, VA, USA
e-mail: ppowell@dvigridsolutions.com; phillip_powell@msn.com

S. McClellan et al. (eds.), *Smart Cities*, DOI 10.1007/978-3-319-59381-4_10

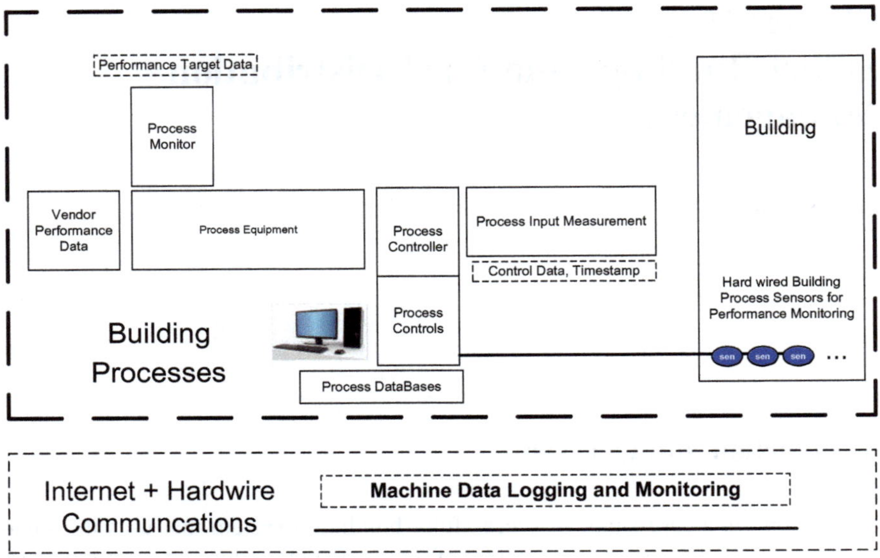

Fig. 10.1 General building components

a failure occurs. Processes are in place to locate the failure and return the failed component to service. Smart building systems actively monitor the key equipment to eliminate the run to failure approach in existing buildings and replace it with a highly reliable and cost-effective smart monitoring system to fingerprint and prevent the failure mechanism.

Electrical Energy Management Systems

Electrical energy management systems comes in all types and varieties. Each has a targeted purpose and the good ones have a clear measurement and verification system to guide operation decisions and measure results. Today's systems are targeted at reducing the demand for the monthly time period that the demand is at its highest. They normally review the power flow into the building and when the peak hour is known, some type of energy reduction(s) is triggered to minimize the electrical usage. As shown in Fig. 10.1 measurements are made on the devices that are to be reduced and the controller will normally turn them off. Simple, directly measured reductions are used to lower demand levels and reduce costs. In addition to direct individual customer load control, customer aggregation using various methods of social media intervention is being tested at this time. Existing systems are controlled independently and the aggregate look at the entire building supply is not normally executed because of the high cost of direct monitoring. Energy efficiency improvements are made by improving the energy conversion process for each independent process and a case-by-case cost analysis implemented to decide to move forward.

A general technology improvement has brought two important characteristics to these types of systems. The first is the Internet of Things communication which provides an inexpensive digital communication system that replaces the direct-wired systems with a network system allowing the machine-to-machine communications. Figure 10.2 shows the addition of this technology and how it generally fits into the multiple electrical processes used in the smart building. The smart building design is about taking this technology and applying it to the electric delivery system in the building and for the first time being able to execute broad demand and energy reductions across all electrical processes with a new method applied to the electrical energy delivery system serving the building. The smart building process is then extended to multiple buildings and coordinated control is executed at a campus level.

Figure 10.3 represents the smart building application applied to two systems that have a targeted purpose (demand and energy reduction) and a clear measurement system documenting their value from reducing the electrical delivery system costs. The target of these systems is to operate the voltage of the buildings and the combined campus at a level that minimizes equipment losses while fully maintaining operation within the voltage design of the equipment. The basic process is to optimize the voltage level by mapping the power flows from the meter points to the loads while learning the power normal process characteristics. Using the derived model the system can continuously optimize the operation at the building or overall campus level while detecting abnormal process conditions of equipment based on learned past operation characteristics.

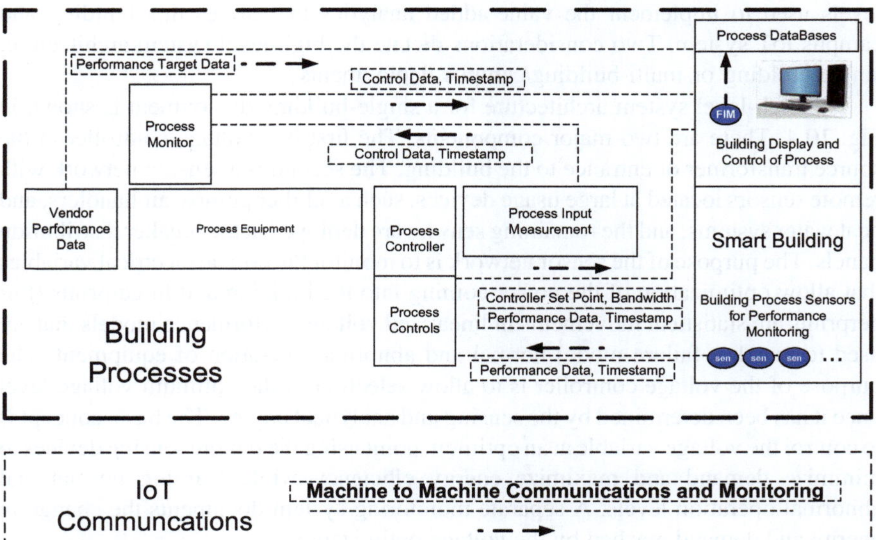

Fig. 10.2 General smart building components with IoT

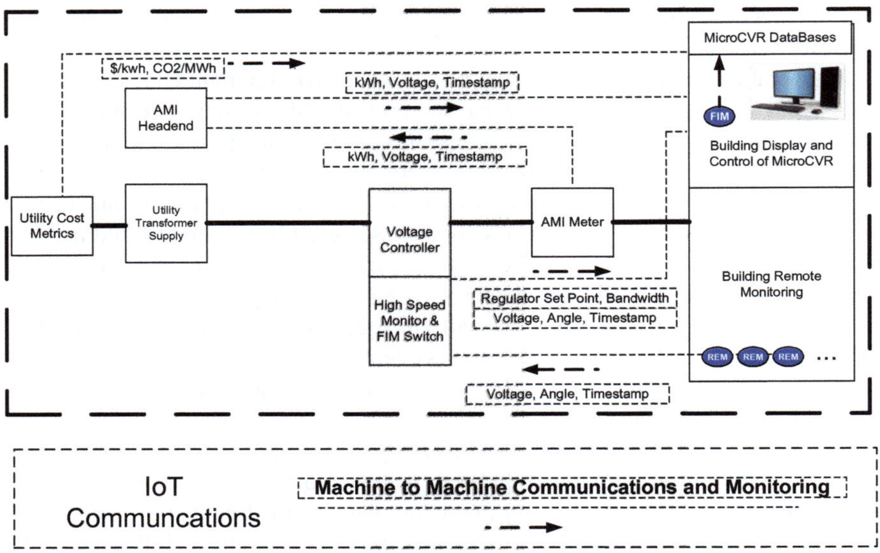

Fig. 10.3 Smart building power components with IoT

Architecture

The next group of sections discusses the overall structure of the Energy IoT architecture along with some of the details of the hardware used to implement this system. The idea is to draw a picture of the hardware level and then the software levels used to implement the value-added analytics that drives this building and campus IoT system. Two considerations dictate the high-level system architecture: single-building or multi-building/campus deployments

The high-level system architecture for a single-building deployment is shown in Fig. 10.4. There are two major components. The first is a voltage controller at the source transformer or entrance to the building. The second is a sensing network with remote sensors located at large usage devices, such as chiller pumps, air handlers, and heat water systems, and the remaining sensors are deployed at the breaker distribution panels. The purpose of the sensor network is to monitor the critical electrical variables that allow optimization of the power coming into the building and fingerprints (fingerprints are statistical movements in linearized voltage performance models that are used to match a failure mode) normal and abnormal operation of equipment. The purpose of the voltage controller is to allow selection of the optimum voltage level once it has been determined by the sensing and analytical engine. The basic concept is to control the voltage variable at an optimum point using the remote sensing devices to minimize demand and maximize energy efficiency while learning normal and abnormal operation levels. A separate monitoring system documents the change in energy and demand reached by the voltage optimization.

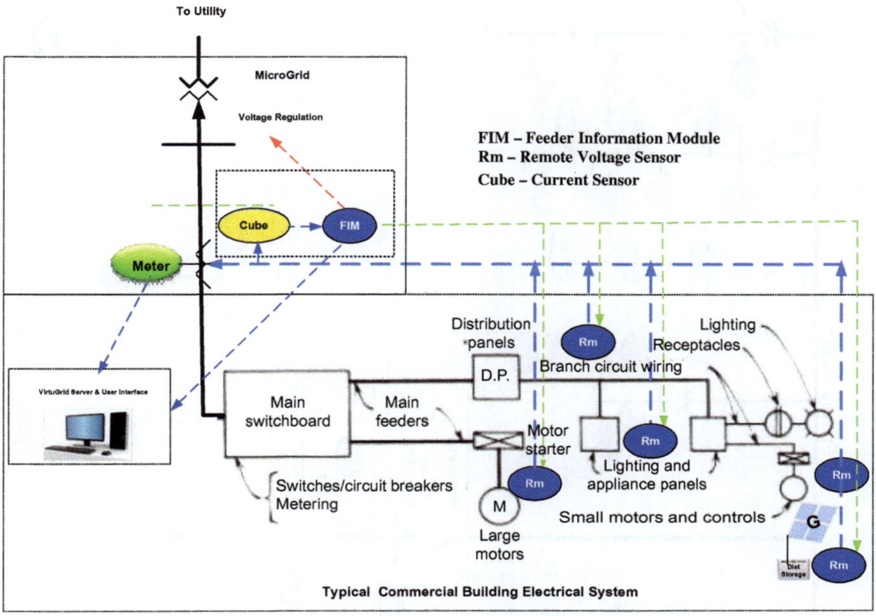

Fig. 10.4 Building sensing for voltage optimized microgrids

The high-level system architecture for a multi-building or campus deployment is shown in Fig. 10.5. There are three major components. The first two are identical to the two major components in the single building architecture (voltage controller and "virtual grid" Monitor System). These two levels deployed at the major buildings on the campus with one or more power feeds to each campus. The slight twist is that multiple small buildings can be aggregated into and overall voltage controller and "virtual grid" system that has its FIM located at the customer transformer(s). The FIM or Feeder Information Module is the receiver of the on-line communications from the remote sensors and the intelligence of the energy monitoring and control system and will be discussed in more detail in the following section. This is shown in Fig. 10.5 with one FIM serving monitor in multiple buildings. This application requires one additional cube receiver for each additional building. The number of remotes is determined by how detailed the optimization needs to be. In general, there are fewer remotes in buildings with a small amount of load and small single device load and more remotes in buildings with larger single load and larger overall loads with multiple load centers.

The third component in the campus style application is an aggregated control system at the total campus level which requires an additional FIM and analytics software to provide a second level of optimization between buildings. In addition, the campus is regulated by a voltage regulation device that the aggregated FIM will control and use the remotes at the building entrances to optimize the delivery from

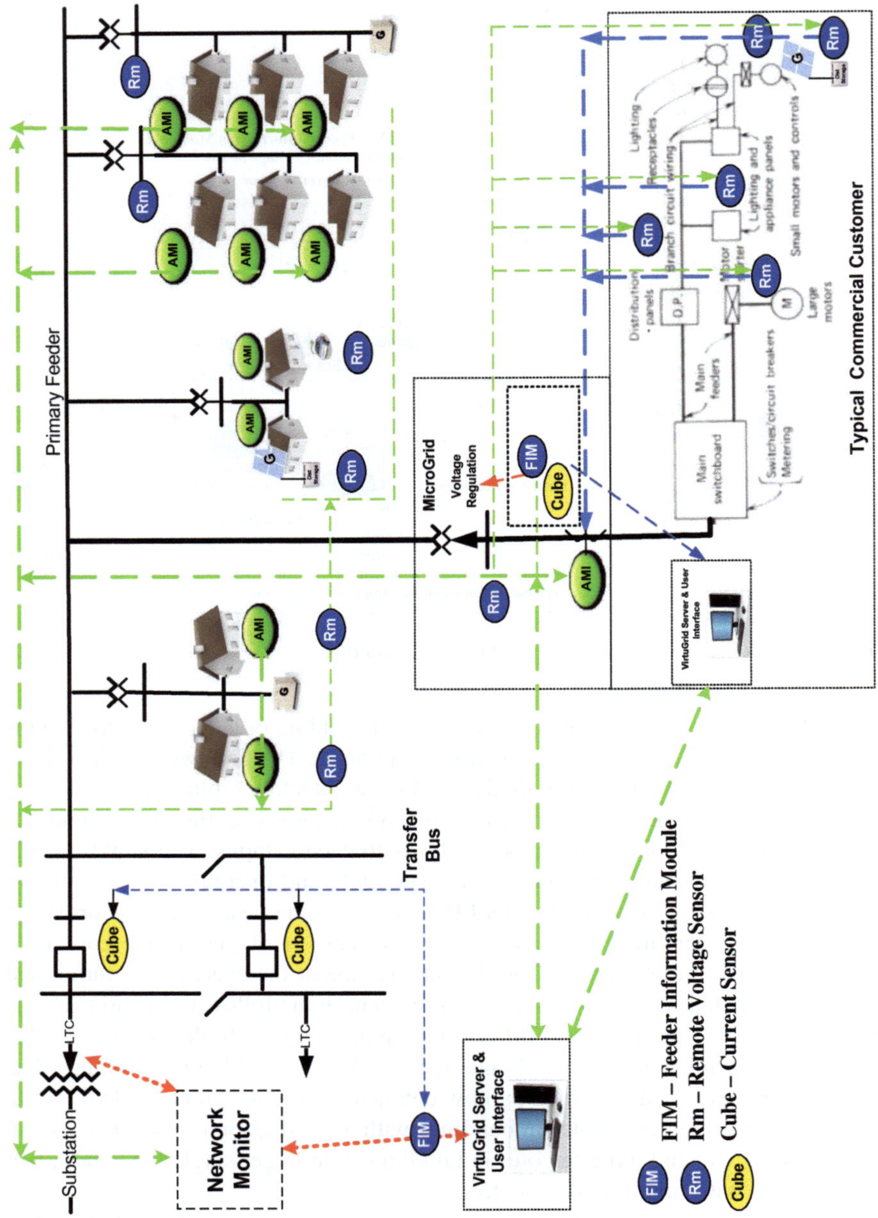

Fig. 10.5 Campus state sensing for building microgrids

the utility sources to the aggregated campus building loads. This aggregated algorithm works on a slower basis to keep from interfering (prevent counterproductive equipment oscillation effects) with the individual smart building controllers which will operate at a faster and more accurate optimization level.

Voltage Controller Considerations

With the general architecture understood I want to dive a little deeper on the primary devices being used in this type of system. The first is the voltage controller. There are two major types. The first type is a slow speed voltage controller which can be a traditional voltage regulator or at a substation transformer level a slow speed load tap changing transformer (LTC). Slow speed means multiple minutes to hours. The second type, a high speed voltage controller which is a relatively new device at the building size power levels, operates in sub-seconds to seconds. The slower voltage regulators are limited in the number of voltage changes they can make before maintenance must be performed on the device as well. High speed voltage controllers do not have a limit on the number of changes in the level before maintenance.

Slow speed voltage regulation is nothing new. The basic technology has been applied on distribution and transmission systems since the early part of the utility development. There have been improvements in magnetic core design and materials as well as production techniques but the general design would be still recognized by someone who manufactured them in the early part of this century. The main characteristic of this type of system is that it is designed to move the voltage slowly to follow the load variations in the primary or high voltage circuit (before the final transformation to customer used voltage levels) while keeping the customer transformer inputs within the voltage range that the equipment is designed to withstand. These mechanical systems are designed to handle a fixed number of operations of the mechanical tap changes before they must be maintained or they will fail. On average utilities are expecting tap changes on a 1–2-h basis to meet their maintenance cycle requirements. These maintenance cycles are usually measured in years between scheduled maintenance. For the maintenance to be completed the device must be bypassed and removed from service for hours to days to complete, depending on the type. Minimizing wear from the number of operations on these regulators/LTC transformers and minimizing costly failure and removal and/or replacement is a major target of the utility operators.

High speed voltage control has been at the utility power level for only a reasonably short time. It developed out of the technologies that were developed for motor control in the 1970s. After they were developed and made reliable for the motor control industry the power levels were greatly increased for applications like DC transmission line converter stations and then scaled back to the lower level distribution and customer applications. The initial development was combined with the power supply development for electronic devices. It was possible to scale these up to higher power levels and apply them to the utility industry.

These are now offered on the competitive market from 100's of kVA to 10's of MVA capability. There are a limited number of suppliers at the high-end range but a significant number of manufacturers at the lower end of the range are working to increase the power capacities of their design. The unique attribute of these devices is that they are sub-second responsive at a minimum and some are much faster than that. They also have the capability to nearly continuously change voltage with no effective "wear" on the hardware systems. It is this capability that has created one of the ways for improved efficiency for the electricity user.

On-Wire Communications [1–3]

One of the unique requirements of this unique smart building application is the use of an on-wire communications link from the remote sensor connection point on the power system back to the location of the cube receiver that is connected to the feeder information module (FIM) for demodulation and processing. This is a unique application that is patented especially for this application. The purpose of the on-wire communication is to geographically and electrically define the position and the connectivity of the device(s) that are being monitored and used for analysis by the remote. Importance of electro-geospatial connectivity is reflected in IEEE 2013–2014 key topic areas. Greater reliance on GIS also supports emergency response decision-making.

The main attribute of the on-wire system is that the bandwidth is limited and forces the designer to limit the amount of information over a set amount of time. It also is very prone to sideband production and coupling of the transmitted information into multiple phases. These characteristics also restrict the bandwidth available. The system carries limited digital data but can send enough information on each packet to communicate the location of the remote sensor, the circuit the remote sensor is attached, and the phase of the circuit the remote sensor is attached. This provides the sensor connectivity information and allows mapping of the system wiring automatically from entrance to load points. It also can communicate the voltage magnitude and the voltage angle at the sensor for all three phases.

Off-Wire Communications

For the on-wire communication, the bandwidth is very limited but does not cost anything to use. On the other side, the off-wire communication has high bandwidths but has a charge for its use. This communication can fall in multiple categories. However, the system is designed to limit cost while providing high bandwidth access to remotes when needed.

The two big reasons for the off-line communications are to provide a timing signal for the on-wire communication and a path for updating the remote software. Because of the attributes of the on-line communication medium, it must use a time division multiple access (TDMA) communication scheme to implement packet communication from the remote sensors to the FIM server and it is implemented only one direction from the remote sensor to the FIM. Any communications from the FIM to the Remotes must be implemented on the off-wire communication path. The on-wire communication implements a type of time sampling of the remote data complete with time stamps, variable values, measurement location, and connectivity to the FIM electrical location. This process of providing this specific information over the wire is referred to as making the location "grid aware" implying that it knows where it is connected. With remotes at the right locations the FIM can make the entire grid "grid aware" and enable the multiple smart building applications that will follow.

The other application implemented over the off-wire communication path is a drill down function. This means that the high bandwidth off-wire path can be used to connect to a small portion of the sensors and do high speed sampling for short periods of time to resolve performance issues or determine why the remote location is operating outside of its targeted optimum band.

The off-wire communications can be implemented on three types of systems depending on the topology of the campus or building that being monitored. For the local building application, an internal LAN can be used to implement off-wire applications. This LAN is a machine-to-machine communication system in each application with the exception the drill down. This makes it a perfect match for the IOT network designs. Once the application moves outside the building to multiple buildings a larger WAN would be needed depending on the distance and number of networks used at the campus site. For large sites that are geographically separated a cell WAN may be employed to implement the full off-wire communications.

Feeder Information Module (FIM)

The FIM is the communication receiver for each remote connected to the power system downstream from its location. It can also be used to house the high-level applications beyond its basic communication role. It is located at a control facility for the single building or a central facility for the campus-level deployment. The FIM hardware is made of sensing cube(s), FIM switch, and FIM processor/server.

The main purpose of the FIM system is to process the TDMA communications from the remotes connected to its monitored power circuits. These communication packets are approximately 4–10 s in length and are sequenced from each remote on a fixed time schedule to effectively scan the network variables providing time-stamped values through the on-wire communications. Also, included in the communications is the angular difference in the variable at the remote and the variable measured at the FIM. In each packet the information for the location and the electrical connectivity is also supplied.

Figure 10.6 lays out the basic hardware application for the remotes and FIM. The design is focused on making installation on existing buildings and campuses easy to implement and provides a method to implement on any size of the network for power delivery. The devices are designed to fit into their location and attach to the network with minimum effort and no power/customer impact. This is essential to making this a cost-effective application.

The auxiliary transformers are the "hookup" point for the FIM receiver to the power system. They are designed to clip around the secondary wires of a power current transformer (CT) circuit with very low burdens that enable connection without removing the power CTs from service making installation very simple with no impact to operation. Up to three auxiliary CTs are connected to the cube which acts as an analog to digital converter (A/D) for the current transformer outputs.

The FIM cube is the A/D converter for the voltage samples using a voltage output from the auxiliary CT which produces a voltage proportional to the current in the power CT secondary. This basically samples the high voltage current using existing power current transformers. The purpose of the cube is to continuously sample the stream of all three CT inputs while staying in time synchronized with the FIM switch, interleave the samples into a data stream, then wait for a timing signal from the FIM switch to ship the samples to the FIM. Up to 12 cubes can be monitored by one FIM by controlling the data stream from the cubes appropriately.

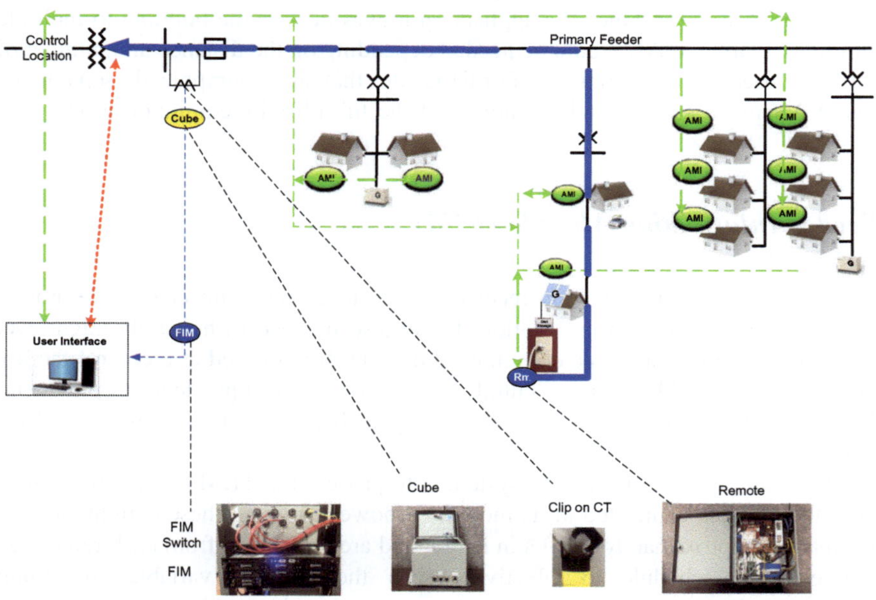

Fig. 10.6 Basic sensing principles and equipment

The FIM switch is the master time controller for the data stream coming from the cubes/circuit currents which contain the data packets from the remotes riding on the power circuits with all harmonics, noise, and fundamental power. The data stream is stored, broken back down to its components and prepared for being processed by the communication demodulation processor.

The FIM software is very flexible and broken into two parts. The first part is the signal processing step which results in demodulation of the signal injected into the power system by the remotes. The second part is the application layer where the information provided by the demodulation process is used to provide information for optimization of the power system being monitored.

The basic communication is centered around using digital signal processing (DSP) techniques to utilize the power conductor networks as a medium to communicate. This medium is already occupied by the fundamental frequency power flows, broadband switching transients, multiple frequency motor drive outputs, harmonic frequency noise, and many other sources that are out of the control of the communication system. This is a limited bandwidth system but it is also one that DSP techniques are readily applicable.

The basic process is that the remote modulates and transmits the signal using one of the TDMA DSP techniques and the FIM samples the power stream to record and demodulate the signal using the same DSP demodulation techniques. Other than the fact that a power system wire medium has very specific characteristics and shares its electrical energy with other applications it is the same as any other medium to communicate over.

Primary Data Collection

Primary data collection is centered around desired cost and deliverables from chosen smart building applications. For this smart building application, the target is operating an energy management system that uses voltage level control to optimize power equipment energy usage and demand based on mapping of power flows through the building(s). This application is driven by voltage measurement. Voltage measurement is one of the lowest cost variables to measure and most available on other equipment making it the best electrical variable to use in the application. Two characteristics of voltage will be needed, because we are operating a three-phase AC power system of complex variables, the voltage magnitude, and angle.

The voltage magnitude is measured at the remote sensors and the FIM for all phases available, converted from analog to digital and coded into the second communication packets that are transmitted on-wire to the FIM. This magnitude quantity is sampled at the remote and used as an input to the loadflow calculation which maps the power flow from the FIM to the Remote locations by converging the loadflow to the measured voltage magnitude values using a stored impedance model of the power system.

The phase angle is very efficiently calculated by aligning the communication starting point with a zero-voltage crossing point at the remote and transmitting the message at the point the magnitude goes to zero. This communication time stamp is sent to the FIM that compares the zero crossing of the remote voltage to the appropriate zero crossing of the local FIM voltage and determines the phase angle shift from the remote to the FIM power bus.

Secondary Data Collection [4]

One of the key application value drivers for this smart building system is its ability to automatically track any changes in geographic information associated with key power system delivery components, limit the amount of manual collection of data for analysis, and to track changes in performance of the power delivery system for the building or the campus. This "secondary data" is based on GPS measurement at the remote location as well as information derived from data analysis and correlation using the primary data measurements.

Normally, it is much too expensive to measure each usage point for all loads. For small loads plugged into the outlets at the usage voltage levels aggregating the measurement of voltage at the breaker panel level is sufficient to represent the physical and electrical characteristics. The approach is to use an aggregate model for the panel loads, such as a ZIP model (Z = constant impedance, I = constant current and P = constant power). For the major loads, such as chill water and heating water systems, a dedicated remote would be placed at the site and a much more device-specific model will be used to represent the system in the loadflow analytics. Physical location of the building will be provided by a dedicated GPS unit at the site and locations within the building for equipment and panels will be provided by floor and wall location information which is stored with the appropriate remote tied to an ID for the remote at the location. This is the starting point for the automated mapping.

Wire pathway containers are the building wire raceways and channels that connect the main equipment to the panel boards and the panel boards to the building GPS point at the meter point for the building where the FIM CT is connected. The pathway containers provide the length of conductors from one device to the other device. A simple algorithm can automatically draw the circuit connectivity from device to panel board to meter point.

Key Analytics and Applications

The system uses primary and secondary data to produce key analytics and to develop important applications. These analytics and applications include geographic mapping, power mapping, and connectivity information. One of the most important capabilities of the system is to map operation in time and power of major equipment power users and the ZIP models of the aggregated loads at the distribution panels. This large use group is usually a small portion of the total number of devices but may represent 60–80% of the power use. These devices have a dedicated remote at their terminals. This provides the power mapping input, detection of voltage variation during device operation, detection of abnormal events on the device, and learning of voltage fingerprints.

The remaining remote sensors are located at the breaker distribution panels for the larger number of devices to be monitored in aggregate. The aggregated devices are represented as ZIP models that have derived quadratic characteristics which are learned from the sample voltage and power data. These aggregate remote sensors also provide power mapping input, detection of voltage variation during device operation, and abnormal fingerprint monitoring.

Connectivity Information

Connectivity information is another application target for this smart building system. In most commercial buildings, there is a set of drawings that show the characteristics of the electrical power delivery system. Their accuracy is a best declining with age of the building. At worst, they have been marked up or portions lost or both and the accuracy of the detailed connections drop at time goes by because of renovation on part of the building and turnover of electrical maintenance personnel who knew the power distribution system by heart.

What is usually determined is that the power conductor paths in a building are limited and can usually be easily documented. The containers of wire path are the starting point for the smart building electrical connectivity mapping. As an example, a quick review of the wiring may indicate that the third-floor wiring is in the ceiling, and runs to a vertical conduit or raceway that returns them to the power distribution point at the main circuit breaker location. This mapping information we refer to as a container. The mapping logic will know that the remote is on the third floor and can determine the connected containers from the remote panel location to the raceway and to the main distribution panel.

Now the connectivity detection system is used to determine which circuit and/or which phase the remote load is connected to by detecting its signal characteristics when the remote transmits the electrical packet in its time slot on the on-wire system communications. Now the complete picture of the power map from the remote is automatically entered into the data file that the loadflow uses to calculate building power flow and uses the connectivity and location information to determine circuit lengths and types. This process makes it very easy to automatically document the connectivity and maintain its accuracy while constantly looking for problem fingerprints in the electrical data. This process is called "virtual grid".

In many cases, the smaller systems that do not warrant a dedicated remote at their location can be monitored by a portable remote that is moved from one location to another. This aids in locating problems in individual breaker circuits, mapping new infrastructure, and does not require a dedicated remote.

Geographic Mapping

Geographic monitoring is the next application to discuss here. What is meant by this is taking the system and appropriately connecting it into an existing geographic information system. These systems are available and the only hardware needed is a commercial GPS unit and some purchased connectivity to systems available at various accuracy and security levels.

Geographic mapping is used on the campus-level smart building level to mark the location of the building and tie the information into a larger GPS view. Once the system is in the building, such as remote monitored loads and panel boards, the location system changes to "third floor, room 122, wall #6" or some other easily identifiable location description that can be translated to a three-dimensional picture of the building facility. The 3D picture is used as the high-level user interface by the operator.

The purpose of the geographic mapping is for the operator to have a quick visual understanding of the status of the building(s) and to drill down on devices to the level desired to optimize the power, detect abnormal fingerprints, and know where and what he is observing on the user interface.

Power Mapping

Power mapping is the high-level analytics targeted to provide multiple benefits to the smart building power system control and monitoring. It rides on top of the information that is described in the previous sections and is enabled to execute a unique application of power mapping. The overall purpose of the application is to provide a power flow with all voltage, current, and power average values for the building power delivery system. The unique part of the application is that the only inputs are total power, voltage, and current into the power delivery system at the supply point and the voltage magnitude and angle information at the delivery point. The input is averaged every 1–5 min using meter accuracy devices on each phase. The voltage and magnitude are measured samples taken in sequence (TDMA communication method) with a communication packet length of 4–10 s in a repeating pattern until all remote data is read.

A single building example of this scheme contains 30 remotes of which 8 of the 30 are on large power usage devices and the rest are panel boards, two per floor, on each floor of the building which has 4 floors. The FIM is receiving communication packets from one remote every 6 s. The total time to receive from all remotes will be 30 time 6 or 180 s. Therefore, it takes 3 min to just scan the remotes. A very important fact is that the remotes are providing the average value of the voltage magnitude for the last one second. This is a sample not a continuous average over the 3 min. What this accomplishes is to make short-term deviations visible that would be filtered out by the averaging process. If it is sampling each second, the operator can connect to the remote through the off-wire system and look at the higher speed one second readings and drill down for just one remote with only a small cost addition.

The power mapping application is ready to execute. It will be used to solve the 15-min power flow information for the smart building. The previous five sample runs from the FIM are loaded into the application. A statistical average of each set of remote readings (five samples per average) becomes the first "guess" of the power flow solution. A standard three-phase loadflow is used to converge the solution to matching input power measured over the 15 min based on meter accuracy readings and first guess voltage magnitudes. Measured differences between the measured voltage magnitudes and the calculated loadflow voltages are used to converge the results to the measurements. Voltage angles are used to provide an approximation of the power factor at the remote load point. This process is run until the best match of the readings is reached. The data is stored and the process is repeated for the next 15 min.

Data Storage

One of the most important steps in the process is efficient and effective storage of the information layer of the combined applications. Again, this is nothing new. Many systems are available to store information but because of the volume and type of information this choice must be carefully selected for cost and usability of the data. The development of the user interface to the data is a critical part of the smart building application. This is where you put the "smart" part of the smart building for the electrical power delivery. It is basically broken into three sections of data that support three levels of application: Topology, Power Mapping, and Statistical Calibration. Figure 10.7 shows the architecture of the data storage system.

Topology is a straight forward database that houses the physical characteristics of the building power delivery system such as floors, equipment locations, building physical characteristics, conductor containers, remote locations, FIM location, Meter point(s), connectivity of power conductors to loads, and characteristics of the equipment and loads at rated values (nameplate, alarm levels, overload levels, …). This information is being continually checked and corrected by the application running on the smart building power monitor system. If a change in the data occurs this is time stamped and the topology is updated automatically.

Power mapping is a little more complicated as described above but it produces a basic loadflow map of the building on set intervals and time marks the solution with a large number of variables such as ambient temperature, cloud cover, humidity, season, time stamps, etc. Only the data necessary to repeat the loadflow analytics is saved along with a reference topology time to allow comparison with future and past operation.

The final database, Linear and Quadratic statistical calibration forms the "smart brain" of the smart building system. This system takes outputs from the loadflow analytics and the measured data from the remotes and uses statistical linearization

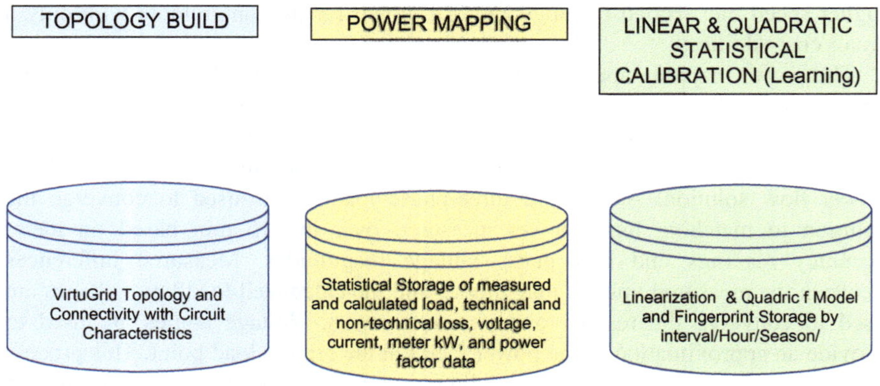

Fig. 10.7 Data storage

and quadratic characterization to fingerprint normal operation and abnormal information that can aid the building operators in early identification of problems detectable in the power/current/voltage variable performance. This system learns the normal and detects the abnormal directing the operator to drill down on potential location in the building of poor operation of equipment or systems.

High-level application run on top of this database to constantly evaluate the performance, guide maintenance, equipment replacement, system performance, and in this case, provides the optimized voltage operating points to maintain the system to minimize demand and maximize energy efficiency.

Energy Analytics: The Engine for IOT Value [5]

The core of the operation of this system is built into the software and analytics of the energy system supplying the building(s) and campus being monitored by the hardware systems discussed previously. For an IoT application the concept is to use a minimal number of sensors, combining them with very powerful applications and analytics to improve the operation of the power delivery by lowering costs and environmental impact.

Once operational this system provides voltage optimization for the building using high speed voltage regulation at the service entrance. In addition, a centralized campus controller using slow speed voltage regulation can provide aggregated building and residential loads, that may be too small individually to warrant a dedicated high speed controller, but can operate with about 60% of the efficiency and demand improvement with this type of control. This is the primary focus of the operating controls which can be automatically or manually tied to the set points of the high speed and slow speed voltage regulation devices. Slow speed regulation can achieve around 4–6% energy reduction and when combined with high speed regulation the system can achieve from 8 to 10% energy reduction. In addition to achieving these efficiency and demand reduction the system uses its software engine to add value by providing the following processes.

Software Engine

The software engine runs in the "virtual grid" user interface server and obtains its data from the FIM on-wire system, the LAN or cell network, and from key internet data sources for processing the information. The system works primarily off measured power voltage magnitudes and measured power voltage angles that are continuously sampled by the on-wire system. These on-wire communications are used to map the wire path from the sensor to the FIM effectively drawing connectivity from the GPS or building location information back across the distribution wire system to the source point where the FIM is located automatically.

GPS locations and building floor and room locations provide the starting point location for the electrical system location. Containers are tracked to locate the wire path and the FIM sensor determines the phase and circuit information using its on-wire communications for each remote effectively enabling the remote sensor to be "grid aware" of its location on the electrical network.

Power mapping is the next step. The geographic and connectivity information is merged into an electrical diagram of the building(s) and Campus geography to automatically draw electrical diagrams from the remote sensor locations to the power source location and combined with the time-stamped meter data at the FIM location is used with a loadflow process to allocate the total meter data to each sensor location using the sampled measurements of voltage. This is executed and values averaged over the 15-min period to provide a "virtual grid" representation of the power flow for the analytics engine.

The linear model engine works to learn the linear characteristics of the power flow calculation and is able to predict the solution to the formal loadflow by using a learned linear model based on key variables such as temperature, humidity, and load using the zip model representation [6]. The linear variables and statistics that describe these variables are stored and used when the "virtual grid" analytics need to forecast variables if samples are not available from the on-wire measurement system. This linear learning also provides the basis for determining when the values measured are out of the normal band and may be a recognizable fingerprint of a problem with the system at the remote sensor.

The Zip model for aggregated load modeling is derived using a quadratic fitting of the constant impedance, constant current, and constant power models. This general representation works very effectively over a short range of power levels used by the piecewise linear model of the power system.

With these models in place the loadflow calculations become the analytics engine matching voltage measurement to allocate the power flow at the FIM source to the power flows at each remote location. This allocation process is solved multiple times and the results are averaged to obtain the average 15 min average readings that become the solutions for the "Virtual Grid." The statistical variations in the power flows and the statistical variations in the voltage samples provide the tools to determine normal and abnormal conditions at the remote sensors. These abnormal fingerprints provide another valuable attribute of the "virtual grid" monitoring system.

Monitoring Benefits [7]

Now that the overall architecture of the "virtual grid" system has been described it is time to summarize the potential benefits of the type of system. The most beneficial value is the savings of energy costs associated with reductions in energy use and demand levels using the voltage conservation principals with slow and high speed voltage regulation controllers. This will be the main target to quantify in the financial section that follows. But in addition to cost savings from energy use, a number of added benefits are produced that have not been financially evaluated.

Use of voltage control to improve efficiency has already been discussed but tracking of voltage variation is a key variable to track when high variable loads and distributed generation are added to a system, especially when these are combined with operating the systems in an independent microgrid where voltage control is more difficult to control and voltage variation is more likely. Remote sensor reports provide the exact information to determine the source of the variations and to engineer solutions that make the microgrid work [8].

Loadflow calculations using linear voltage allocation techniques are the core analytical engine for the "virtual grid" analytics. It is important to note that the loadflow is a full three-phase model with all three-phase variables measure and calculated and with the full neutral path model. This provides a clear method to track individual component equipment ratings and measurements at every level of the power flow. In addition unbalance performance is directly calculated by this system and can be statistically tracked to provide tweaks to the operation that will aid in operation of the individual high speed single-phase voltage regulation providing even more improvements are power optimization [9].

If an outage of a system results in an outage of a remote, it stops transmitting and allows the detection of an outage event at the location. This is obviously an abnormal fingerprint reported to the user interface. The extent of the outage can be determined by the number of remote sensors lost and this is detected in each communication cycle. Immediate knowledge of the outage and extent of outage will greatly improve the response time and the restoration of the lost system components. Statistical tracking of these types of events will enable targeted improvements for fault-prone equipment.

The optimization of voltage across the network will normally occur at the bottom of the equipment voltage band. Operating equipment in the lower part of the equipment voltage rating generally reduces heating and voltage stress on the insulation. This has been documented to make insulation last longer and extend the life of the equipment [10].

The detection of abnormal operation of equipment is a major benefit of this type of system. One example is detection of a building air conditioning system abnormal operation from loss of coolant. The normal fingerprint of loss of coolant is that the system loads for the compressors will usually decrease consistently for each running level. A statistical shift in load level for this system could be used to detect the loss of coolant pressure below required level. The air handler systems work in an opposite way. Dirt and filter contamination may cause the fans driving the process to load to a higher level. Detection of a statistical load variation above the normal could be used to detect this condition. The analytics engine can use voltage and power characteristics and learn the normal operation and detect the abnormal operation from a very small number of variables [11, 12].

The last item is power quality. The detection of the on-wire communication requires high speed sampling of current and voltage values at the FIM location. An inexpensive routine that takes this sample data to a power quality engine and determines the harmonic and noise characteristics are applied to continuously track the normal level of power quality for the site and can be used to detect abnormal power quality operation that could be caused by damaged filers, poorly operating power equipment or abnormal load operation [13].

Economics of Smart Building IOT 101 [14]

It may seem strange that I am placing economics under operations for the smart building technology example. But the reality is that anyone designing an IOT application with smart building applications must have one eye on the technology and another on the financials of savings, development, and deployment. They cannot be separated. Entrepreneurs first mistake is to concentrate 100% of their activity on technical development which they love and ignore the economics. But economics is the required but essential root canal of developing a successful system. Make it quick and simple to limit the pain but never avoid taking care of it. When this is not done, statistics show that most great ideas are successfully deployed by the second company that buys the idea at a discounted price because the entrepreneur was unable to reach sound economics.

To avoid this type of outcome, this section will show a minimal method to track the required economics for this type of smart electric building IOT system. There are many methods of tracking business costs and benefits and making sure they are staying at a reasonable level but the object here is to provide a quick rule of thumb that provides reasonable assessment with the least amount of effort. To accomplish this a simple present worth method will be applied which can be used to track and should be used for verification after the development and deployment are completed. The process of development and deployment will be broken into three phases. The first is the research period, second is the demonstration/pilot implementation, and the third is the production deployment. It is time to get specifics for the smart building electrical system technology. First, the general economics of electrical demand reduction and electrical energy reduction is reviewed. Second, the smart building IoT architecture economics are added. Finally, the application of the simple present worth analysis is applied for economic evaluation and tracking for each of the three phases of system research, development, and deployment.

Simple Present Worth Analysis Method

Using the present worth method (PWM) for economic evaluation and validation is a direct method of measuring performance of economics for a system over a long-time reflected back to the present day. This accounts for a lot of the variables of cost and benefit and places a repeatable process around them for evaluation and tracking. Figures 10.8 and 10.9 graphically show the inputs for the application of PWM to the smart building IOT systems. PWM is used in this case to put the cash expenditures of a development business on a present worth basis to look at the impact of various steps of development on the business cost and revenue structure. This analysis can be thought of as representing an investment inside a company on a new technology line or the evaluation of a new independent business where the

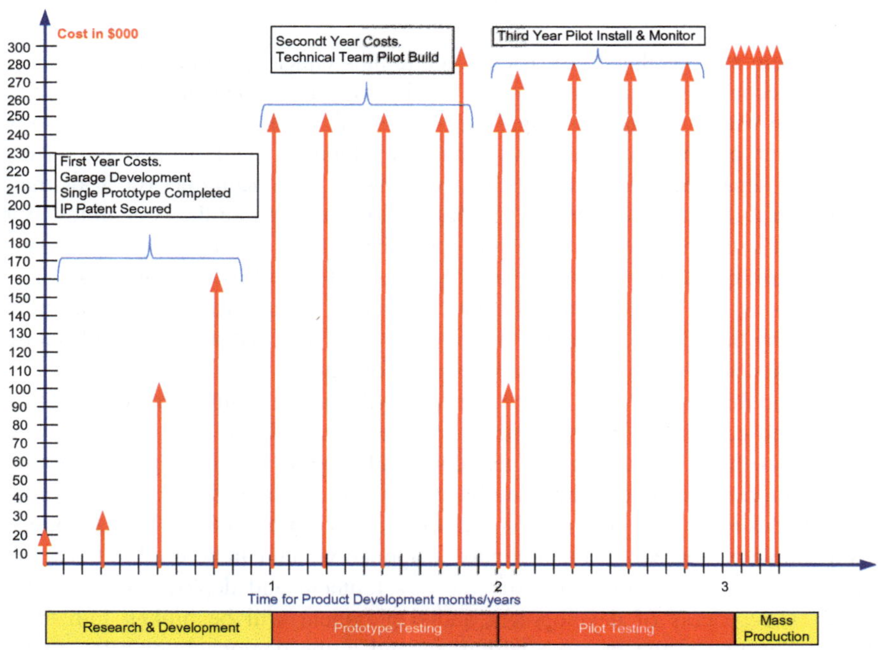

Fig. 10.8 Research, development, and piloting costs

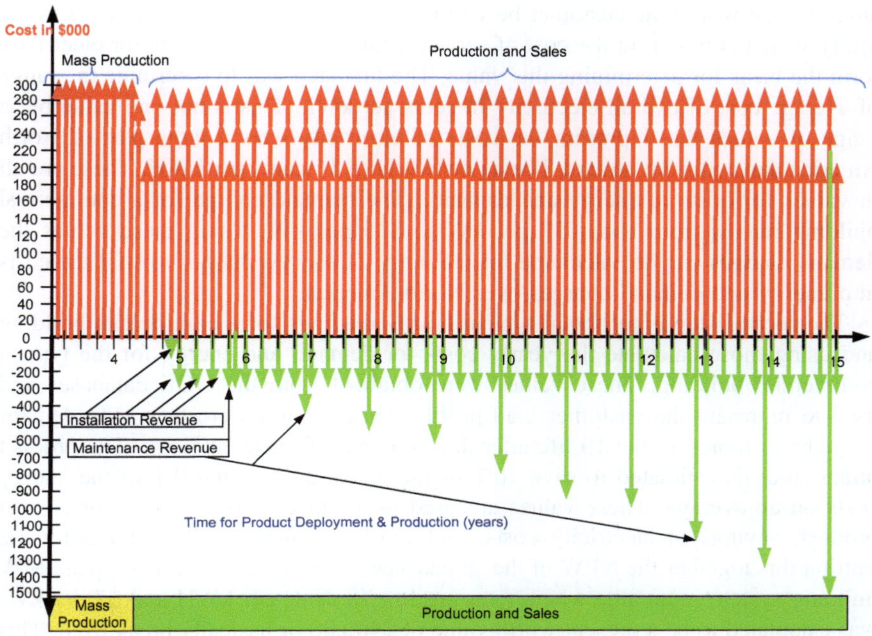

Fig. 10.9 Production costs and revenues

new process is the basis for forming the new business. In either case, we need to know whether this idea can reach a point of paying for itself and the required performance for meeting the needs of the company and investors.

The product build process is broken into three major phases, research, development, and production. Within these two categories are steps that are included in most development projects. This section will discuss how these are applied to a typical development of a Smart City application. The example is centered on the electrical savings from reduction of electrical energy demand and improvements in electrical efficiency discussed in earlier sections. There are two value drivers that must be evaluated against the cost of the system and each kept in site as the development and production process executes. The first is the value of the service and the second is a cost of providing the service.

Economics of Demand and Energy Efficiency

This smart building technology provides an improvement by increasing energy efficiency for the building (kWH) and decreasing the electrical demand (kW). The first question becomes is how much is this worth to the customer and based on that answer what would the customer be willing to pay. What the customer pays to the utility which is based on the cost of service economics and to tariffs for electricity form the basis for determining this value. The first step was to determine the range of average cost of overall electricity in cents per kWh. This is not one value but a range of values that represent the costs of electricity over a large part of North America that run from around 8 cents per kWH to 16 cents per kWH. These are all in values for costs of energy and capacity. The demand charge for a commercial building ranges from about $7 per kW to $15 per kW. Using these values the demand charges can be calculated for the range of energy charges to split the costs into energy and demand charges over North America.

The second step is to obtain typical hourly readings for a commercial building and using those calculate the yearly costs for demand and energy for the typical commercial building. These values were obtained from the DOE databases and used to represent the customer load profile. These values were then extended for 10 years to represent the 10 life after development of the technology. The product under study is estimated to save 10% of the demand costs and 9% of the energy costs on an average. These values are used to produce a 10-year view of cost of product, savings in electricity costs, and yearly maintenance fees for software. Putting this together the NPW of the spreadsheet over 10 years for a 1–2 peak MW building system in an area where rates are low (8 cents per kWH and $7 per kW) was calculated to be a present worth value of $400.00 or basically break even. This means that the system pays for itself over the 10-year period using a 10% discount rate. The other calculation for higher rate areas (15 cents per kWH and $15 per kW) result is a present worth of $400,000 based on the per unit costs of $300,000, install costs of $40,000, and a yearly software maintenance cost of $17,000 per building.

Economics of IOT Hardware/Software Design

With the first financial step done the second step is to take the results from the first step and evaluate if the price determined by the value created for the customer is enough to drive a business that produces a product. In evaluating these two steps it may be necessary to loop through them multiple times until a business financial plan and customer value balance is reached that support a positive sales basis.

It has been determined in step 1 that the customer will find value in providing this 9–10% reduction in energy cost product with an initial payment of $300,000 per building and an additional $40,000 installation cost up front. The on-going software maintenance cost will be approximately $17,000 per year per building installation. These values are put first into the yearly revenue stream multiplied by the number of required sales per year in the 10-year production cycle.

Figure 10.8 costs are then rolled into the spreadsheet representing the original research costs, the build of the first prototype costs, and the deployment of the first pilot system costs shown in the first 3 years. And just note that there is no offsetting income from this period. In the fourth year is a required process of converting the pilot and prototype to a mass production system to reduce the remote sensor costs usually involving conversion to a field-programmable gate array (FPGA)/ application-specific integrated circuit (ASIC) technology and acquiring the required safety standard certifications. This is usually about half of the production costs for the product. When this is put into the PWA in first year value based on 10% discount rate this represents a cost of a little over $5 million in research, development, and piloting before the product hits the market.

The next step is to build out the cost structure. Figure 10.9 generally lays out the basic costs for the production and business. Using the customer value information from the first step the amount charged for the system to the customer is derived in a cash stream based on the number of sales per year. Then the costs from the development process and the installation and maintenance are layered into the cost stream. Then comes the big assumption, that the owner(s) can find a bank to payroll the costs until the business is on its feet. The first cut will be done with a 10% cost of capital just to see where it calculates out. Then the present worth analysis is applied to the end of year cash for the company assuming it is being funded by the capital process at the low-cost rate of 10%. After putting this together, the revenue cash is put in the PWA and surprisingly it has a net present worth of $36 million based on selling 8 systems per year. The true test is the net cash PWA over the 10-year period used to measure the value of the business. This is a disappointing negative $6 million. This is a good indication that you will be losing money after 10 years and is not a good plan.

The next question is what does it take in sales to breakeven after 10 years. This is calculated by increasing the number of sales per year from 8 to 14 per year. This will push the sales process but is possible and results in a present worth of approximately $2 million after the 10 years. But with the low performance the bank now says that it can only fund $1 million of the development and you must get the other funding from sources charging 15% instead of 10%. Back to the present worth analysis, a group of loans are created to finance the remainder of the investment cost for the project. Placing the loan payments at 15% into the cash stream and re-evaluating the present worth increases the cost again and it is necessary to sell 24 systems per year instead of 14 and the 10-year present worth is around $5 million to pay the investment loans and the project development.

One last financial issue. Someone gives you the option of supplying the funding for the development but instead of a loan structure they want 20% of the company value for supplying the development money. Again, the present worth analysis is used taking 20% of the net cash and allocating it to the investor but setting the loan structure as a no interest loan. This will roughly represent the transaction in present worth analysis. This is a big impact requiring a sale of 60 units per year to create a present worth of approximately $3 million.

The main purpose of this section is to know the value drivers to the development and business being undertaken. As product targets are found the analysis gives a very good indication of whether this sales direction can be beneficial to the business and the customer. With this completed the next step is to look at target systems.

Customer and Business Economics

The economics of the system build is only one dimension of the economic problem faced by the inventors of the IoT system. The most important part is what is the value of the system to the customer. This customer value must be communicated clearly and result in some of the value going to the product supplier and the remaining going to the customer who is willing to invest in the application.

Plugging in the residential type loads into the PWM results in a quick recognition that the technology is too expensive for the small loads used by residential customers. The main barrier is the amount of load and savings produced by the residence. The analysis quickly determines that around 1–2 MW of peak load is needed to justify the cost structure. With residential this takes an aggregation process and greatly increases the number of remote sensors. It takes about 150–300 residences to achieve the type of load level. It means that aggregation of residential customers using a slow speed voltage controller is possible but not direct high speed control.

The single smart building system is a good target and just meets the economics for the PWM analysis based on value and production costs for the lower cost electricity levels and more that make the target at the higher end of the electricity costs. They form a clear target for the smart building electric technology. Slight increases in cost can easily impact this economic model and care has to be taken to staying within the business plan for initial development and for keeping costs per system in the ranges projected.

The campus system takes the form of a combination of the two systems above with the unique attribute that the overall campus is on a slow speed voltage regulation system with a lower savings level and the appropriate buildings are on the high speed system, combining the two makes the best sense for the long-term market of the business. The campus system has the best ability to meet and exceed the present work analysis and should be a key focus of the sales for the system. Customer value and business value are maximized by doing this. From a financial perspective, the remaining task is to estimate the market of campuses and buildings targeted and set the goals to meet the number of sales per year on a production, installation, and operation basis.

Cities are made up of residential, commercial, and campus types of loads for a very large percentage of the electrical energy use. Aggregation of the concepts from individual building, campus aggregated buildings, and aggregated residential and commercial in a campus layout is a very powerful tool that covers a major part of the Smart City electrical infrastructure.

"Virtual Grid" Development Example

This section will discuss the actual implementation of the unique technology and its history since the garage developed concept began in 2010. The first step was an actual test of the system completed from the author of this section's garage approximately 5 miles to an electrical substation to test on-wire communication in 2010. This test was being completed to prove to the investors that the technology would work on a practical electrical supply system from the residential customer to the substation where the equivalent of a feeder information module (FIM) was located. After two test periods over a month long period the system was successfully tested using a very crude test setup. The next step was to develop a true pilot level system to deploy in a building at a local university.

Virginia Commonwealth University Campus Network

Virginia Commonwealth University (VCU) volunteered their four-storey engineering building to be the first building to implement a version of the control system [15–17]. This one was a true demonstration prototype. Thirty remote sensors were deployed across the building footprint including the major chiller and compressor equipment in the basement as shown in Fig. 10.10. The large remote

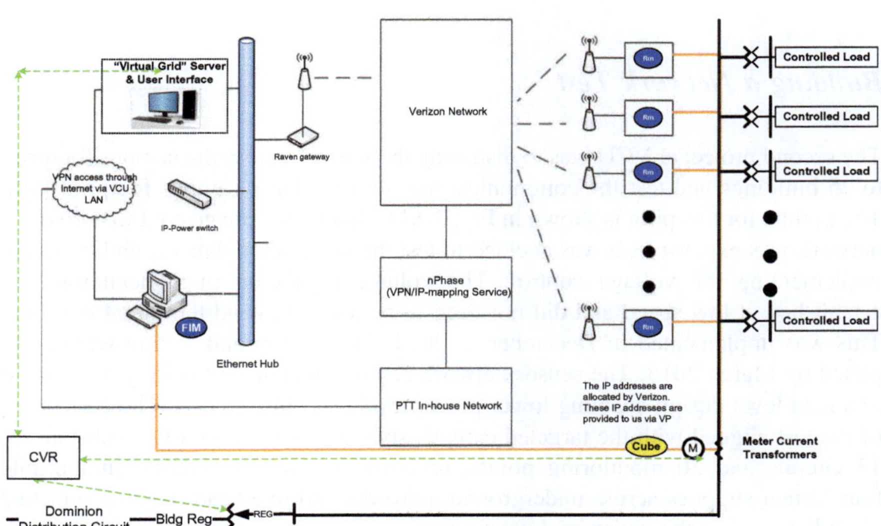

Fig. 10.10 Smart building slow speed "Virtual Grid" engineering building

sensors contained dual large power supplies for executing the on-wire communication and cell network communications for the off-wire network. The FIM was connected at the entrance to the power supply in the basement and the voltage regulation devices were single-phase voltage regulators located on the high voltage side of the service transformer. After about three months of operation and testing, the reliability of the measurement and control system was completed and the application testing began.

The performance of the system was documented to be a saving of approximately 3% for voltage control to the first step in the voltage reduction. The second step would allow almost doubling of the first step results. The movement from slow speed control to high speed control was estimated to be between 6 and 8% based on the results of this first slow speed testing. The communication networks demonstrated reliable and successful operation as described in the previous sections. Overall the test was successfully completed with 30 sensors, one FIM and one "virtual grid" Server. The cost was within the target band and included the slow speed voltage regulator. The prototype technology was discrete and was the main cost in the project.

Low speed voltage regulation affected the results. The starting and stopping of process had to be anticipated by the voltage regulator setting because it was not able to respond to the quick voltage drops from fast process load increases. This meant that almost 40% of the savings was lost because of the regulator speed. The concept of using a high speed controller to execute of this grew out of this observation. But the high speed regulation was a new technology at these types of power levels and only a few vendors could handle the building size loads necessary to make the product cost-effective.

Building a Network Test

The second project at VCU was to distribute the sensors across the campus footprint to 25 buildings and test the communications across a large campus footprint [18]. The outline for this pilot is shown in Fig. 10.11. Since the voltage control across the network was expensive, it was decided to test the application data capability before implementing the voltage control. The voltage regulation implementation was established at low speed and did not need to be tested beyond the initial building. This was implemented in December of 2014. The startup and testing were completed by March 2015. The sensor network communication technology was turned to a next level design enabling lower power levels and lower costs. The geography of the test aligned with the targeted campus style application with two substations, 13 circuits and 30 monitoring points, reporting to two substations on multiple transformer supplies across underground networks and overhead facilities attached to chillers across the entire campus network.

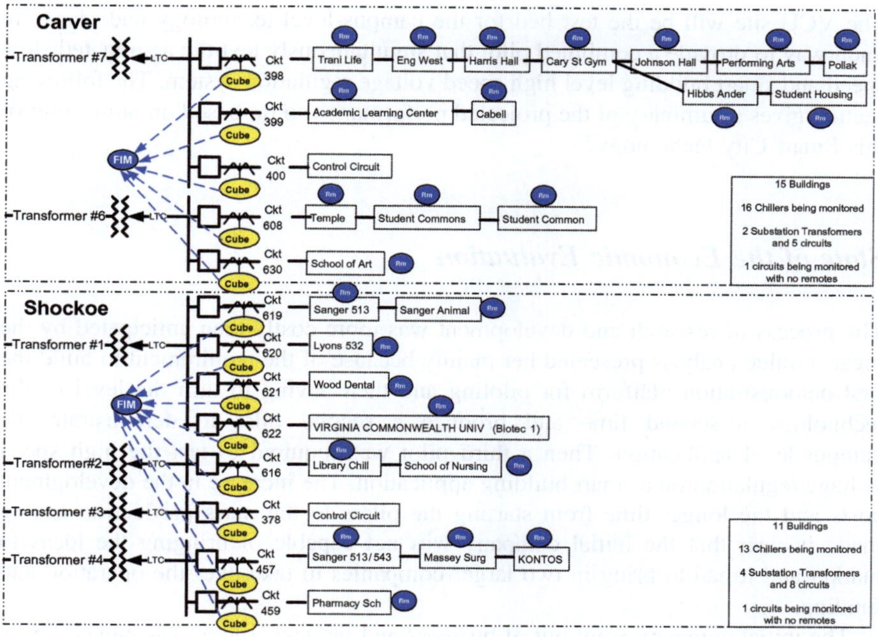

Fig. 10.11 VCU Chiller "Virtual Grid" project

In conjunction with this testing, a second building at another location was chosen to test the high speed voltage regulation system. This application was approximately 1 MW of peak load with 12 remote sensors and high speed voltage regulation device. This installation was completed in August 2016 with 12 building sensors and a FIM switch located at the high speed voltage regulator and the FIM computer located in a separate control building with the "virtual grid". This application combined with the VCU distributed communication network across the campus tested the remaining concepts to enable a evaluation of a full demonstration and pilot of the overall "virtual grid" platform.

This testing is still on-going. The campus communication system is working and plans for full reliability testing are underway. The technology is performing well so far. The second test site for the high speed voltage regulation is demonstrating savings level above where the projections predicted. The savings for this type of control is looking as if it will provide between 9 and 10% energy savings when in place. These results are encouraging and we will continue to run the demonstrations until full reliability testing is complete.

We expect this final piloting to be completed by August 2017. The next step is to detail the line of site to the production level remote sensors. This is a requirement to meet the rigorous cost requirements in the financial plan. The FIM/VirguGrid development is much less costly and will be done in conjunction with this effort.

The VCU site will be the test bed for the campus-level technology and plans are underway to provide a combined pilot that simultaneously text the aggregated slow speed and smart building level high speed voltage regulation system. The following section gives a summary of the projected performance so far based on our testing of this Smart City technology.

State of the Economic Evaluation

The process of research and development was more costly than anticipated by the present value analysis presented her mainly because of the requirement to build the first demonstration platform for piloting and then having to turn the level of the technology a second time and produce a second pilot to demonstrate the campus-level application. Then a third pilot was required to test the high speed voltage regulation on a smart building application. The increase in the development costs and the longer time from starting the pilots to the mass production testing made it clear that the initial company was not capable of bringing the ideas to market and it had to bring in two larger companies to take over the operation and funding.

The initial company went out of business and the technology was picked up by the two companies to reset targets and focus the mass production process to meet the financial targets of the businesses and developers. The following results were determined in the initial research and development.

Using a slow speed voltage controller with the "virtual grid" system was not cost-effective when applied at the single building level. Multiple small buildings could be aggregated into a campus control and the slow speed voltage regulation could be effectively applied with saving in the 2.5 to 3% range for aggregated campus power levels exceeding 50 MW.

Smart buildings with power levels greater than 1 MW and lower level electricity costs could just meet the financial models and be cost-effective. This was dependent on the peak to average load ratio of the building. Economic case studies could easily determine if the building loads would produce a successful financial operating level to pay for the "virtual grid" installation over a 10-year period.

The best application found so far for the application is campus-level "virtual grid" deployment with slow voltage regulation at the campus level and with smart buildings that meet the 1 MW level of load with high speed voltage regulation combined into a very efficient and cost-effective application for energy control with the other benefits of the "virtual grid" application. This smart campus energy supply system appears to exceed the performance requirements projected for excellent financial performance with this smart cities IoT technology application.

Summary

These multiple pilot applications have demonstrated the effectiveness of using the value drivers of energy efficiency and demand control of electric power delivery and combining them with high and slow speed voltage regulation with the "virtual grid" technology to produce a valuable smart cities application. This system can significantly improve electrical usage and reduce the impact of providing energy to the city infrastructure and demonstrate the ability to produce an application that improves the quality of life in a cost-effective manner.

References

1. Loporto JJ, McClellan S (2013) Intelligent power system and methods for its application. USA Patent No. 8519832, 2013
2. Bernheim HF, Martin MR, Berens SJ, Loporto JJ, Niemann TV (2015) System and method for grid based cyber security. USA Patent No. 9059842, 2015
3. Martin MR, Bernheim HF, Abendschein AL (2016) System and methods for synchronizing edge devices on channels without carrier sense. USA Patent No. 9380545, 2016
4. Bernheim HF, Hansell J, Martin MR, Bullard JM (2016) System and method for inferring schematic relationships between load points and service transformers. USA Patent No. 9438312, 2016
5. Reklaitis GV, Ravindran A, Ragsdell KM (1983) Engineering optimization. Wiley-Interscience Publications, New York
6. Kailath T (1980) Linear systems. Printice-Hall Inc., Englewood Cliffs
7. McDonald JD et al (2013) Smart grids infrastructure, technology, and solutions. CRC Press, New York
8. Powell PW, Tyler SJ, Peskin MA (2016) Maximizing of energy delivery system compatibility with voltage optimization using AMI-based data control and analysis. USA Patent No. 9367075, 2016
9. Powell PW (1981) Modeling of substation control circuit transients due to lightning and switching surges. Maters Thesis, Purdue University, West Lafayette, Indiana
10. Harlow JH et al (2012) Electric power transformer engineering. CRC Press, New York
11. Peskin MA, Powell PW (2016) Electric power system control with planning of energy demand and energy efficiency using AMI-based data analysis. USA Patent No. 9354641, 2016
12. Winston PH (1992) Artificial intelligence, 3rd edn. Addison-Wesley Publishing Company, Reading
13. Heydt GT (1991) Electric power quality. Stars in a Circle Publications, West Lafayette
14. Baughman ML et al (1991) Engineering economic analysis: overview and current applications. IEEE Service Center, Piscataway
15. Tyler SJ, Powell PW (2016) Management of energy demand and energy efficiency savings from voltage optimization on electric power systems using AMI-based data analysis. USA Patent No. 9325174, 2016
16. Powell PW, Parker SK, Bollbach MA, Pruett ML (2013) Voltage conservation using advanced metering infrastructure and substation centralized voltage control. USA Patent No. 8577510, 2013

17. Powell PW, Parker SK, Bollbach MA, Pruett ML (2013) Voltage conservation using advanced metering infrastructure and substation centralized voltage control. USA Patent No. 8437883, 2013
18. Powell PW (2016) Secondary distribution grid information system. USA Pending Patent Filed December 30, 2016

Chapter 11
Smart City Lighting

Chris Boissevain

Introduction

Smart City Lighting is evolving rapidly because of the confluence of multiple technological revolutions. Light is going digital through the light source revolution of Light Emitting Diodes (LEDs). LEDs are rapidly replacing the shorter lived, less efficient and only crudely controllable legacy light sources of high intensity discharge metal halide and sodium, fluorescent and incandescent lamps. Haitz's law [1], which forecasts the improvement in LED technology, has been on track for 50 years and is projected to continue unabated for years to come. The efficiency and cost curves of LED systems surpassed the best legacy lighting a few years ago (Fig. 11.1).

Also, and as any visitor to SXSW Interactive [2] over the last few years would know, the Internet of Things (IoT) and artificial intelligence (AI) are the other ongoing revolutions. The use of cell phones, data storage in the cloud, embedded intelligence and "Big Data" from better, cheaper sensors are beginning to achieve unprecedented control, responsiveness, and efficiency in lighting through characteristics that were previously nonexistent in legacy lighting systems, such as dimming and hue change.

New developments in lighting systems are also enabling unexpected new ways of using light for communication, wayfinding, as well as cultural, artistic and civic expression and branding. These applications are available because of color control, various levels of resolution, and the urban scale of controllable pixels, as well as high-bandwidth communication using LiFi [3].

C. Boissevain (✉)
Philips Lighting, San Marcos, TX 78666, USA
e-mail: chrisboiss@gmail.com

S. McClellan et al. (eds.), *Smart Cities*, DOI 10.1007/978-3-319-59381-4_11

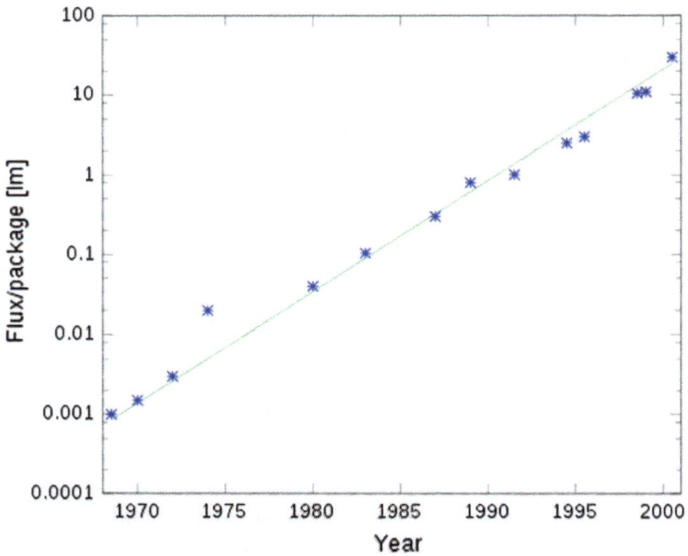

Fig. 11.1 Haitz's law and actual LED flux results over time [1]

Urban Smart City Lighting Vision, 2050

Cities are projected to contain 66% of the world's population by 2050 [4], and megalopolises or "super cities" will be more numerous. Cities will be more interconnected and much more sustainable. Just as there is more glass than ever, cities have more transparent governance, and network technology and digital adoption by the entire population enable a more direct, informed and vital democracy.

The city will become one big computer and citizen/government/services network, where every luminaire will be a display and sensor to the interactive AI that controls and responds to the citizen's needs of safety, communication, and even personal expression. Powered by solar, wind geothermal, and fusion, our civilization will be rapidly approaching "Kardashev Type 1" or "planetary civilization" stage, being able to harness most of the energy received from the sun on the planet.

Transnational, citizen-centric democracy will be facilitated by real-time global connectivity, communication, and a global urban culture of problem-solving. In this civilization, a role of lighting will be to visually connect the cities within themselves and to other cities with interactive sensors and digital display. Lighting fixtures will be available in a large range of resolutions and locations, with graphic, decorative and motion lighting that overlays the city architecture and surfaces with variable transparency. This web of digital light will be highly utilitarian and low glare, providing wayfinding for emergency services, locators for self-driving cars, extra light for rescue services.

Lighting will also generate new and unexpected opportunities for urban identity and place-making, with interactive and animated night scenes celebrating and expressing the values that a city and its citizens wish to convey to themselves and to visitors. Safe and beautifully lit cities will be covered by stars in the night sky due to highly granular light distribution and dimming control. Flashing digital near-ultra-violet (UV) light will protect migrating birds and other animals from collisions with buildings, transport, and windmills. Nature and habitat will be woven into the urban structure, like the hanging gardens of Babylon, as in Fig. 11.2 [5].

As a way to avoid the collapse of the megalopolis described by Geoffrey West [6], restored parks and rivers will divide cities via green belts, like cells splitting in mitosis. Indoor, artificially lit and centrally located parking garages will become disused because of self-driving cars and better public transport, and rooftop farms will generate fresh organic produce and fish and even insect protein. These oases will be scattered throughout the megalopolis, and will counteract the relentless disappearance of arable land.

Creativity will be unleashed by widely available technology and the nearly free energy available due to previous capital investment in renewable energy infrastructure. Lighting controlled by artificial intelligence will provide expanding opportunities for public art forms emerging from software-mediated interactivity with environment, actions and information, and individual, collective and orchestrated vision.

Fig. 11.2 The hanging gardens of Babylon, an ancient mix of urban architecture and nature [5]

Elements that Could Enable This Future

Several technological, structural, and environmental factors are important in enabling the future of smart lighting. Factors such as Haitz's Law [1] for lighting and Moore's Law for computer chips will play central roles in the future of smart lighting. LEDs, cloud computing, and embedded sensors will continue to become more powerful, cheap and ubiquitous. The data these devices produce and that is subsequently analyzed by humans or AI "lighting bots" will generate tremendous sociological insights.

Additionally, digital controls of aspects of light will continue to penetrate the market and provide improved quality of life for humans, as well as wildlife. Concepts such as spectral power distribution, (the mix of different wavelengths of light) and Color Rendering Index, (how well light helps our eyes see colors) will become embedded in commonplace items to improve color response, save energy, and improve perception.

Technologies such as the patented Philips ClearField and ClearSky systems, which tailor the light to be much less fatally attractive for birds, bats, turtles, and amphibians, will become increasingly important for wildlife management. Lighting that exploits the quad-chromatic vision of birds, (most birds can perceive ultraviolet light with an extra cone cell), pulse radiation, seasonal biologically tuned lighting levels and patterns will also play important roles in the management of wildlife as humans continue to encroach on typical habitats. The simple conscious use and placement of light and windows based on LEED recommendations [7] on buildings, towers, and other structures will also slow the accelerating species loss.

Technology penetration into unconventional applications will also increase and become pervasive. Innovations such as distributed lighting embedded in building materials, facades, and roads like those developed by Solar Roadways [8, 9], and in the form of point source, directed light, and glowing surfaces like the glowing bike paths in Lidzbark Warmiński, northern Poland [10] will no longer be unusual. For example, LEDs are already powering a trend of embedded lighting in all sorts of building materials such as concrete, wall board [11], and glass. An impressive application of these concepts is already visible on the Yas Hotel in Abu Dhabi, where the architectural surface glows with embedded LEDs [12].

Technology can make it weird too. People may become mobile urban lighting themselves; with genetic engineering to allow our children to glow with phosphorescence like genetically engineered, glowing pigs [13], or use conductive tattoo ink and glue LEDs to their skin [14] Some already go a bit too far, and embed LEDs surgically under their skin for that eccentric cyborg-effect [15]. Don't try this at home!

Lighting is already embedded in clothing usually for prom costumes [16], and music acts. It is also used in Motorcycle and Bicycle riding jackets and helmets for safety. Later this could be networked with cars for more robot awareness.

Agricultural activities will not be immune from the effects and benefits of smart lighting. Efficient solar/wind powered hydroponic agriculture, lit with LEDs whose spectral power distribution is tailored to each crop species will become commonplace. Exploiting the disused urban architecture spaces, leveraging empty car garages, using converted shipping containers, etc., will drive novel uses of smart lighting to invigorate "indoor farms." Agriculture will become interspersed throughout the megalopolis, along with aquaculture [17, 18], all aiding with city air purification and oxygenation, and carbon sequestration. This approach will also be popular for countering urban desertification through increasing humidity and lowering roof reflectance. All of these fundamental human endeavors will be enabled via smart lighting in the Smart City of the future.

Additional factors involving, leveraging, or becoming enabled by smart lighting might include:

- Network densification provided by streetlight and various luminaire-mounted transceivers, LiFi emitting luminaires, WiFi as a universal urban utility, 5G and better cellular.
- Windows that generate solar power and emit low glare light or reduce glare and solar gain, switch on privacy/transparency, work as various sized graphic light pixels for "architainment" and bird strike prevention.
- Luminaires that are responsive and data gathering with sensors of occupancy, daylight, moonlight, security, air quality, noise, ambient light levels, Color Correlated Temperature (different nuances of white, from warm to cool), for health, mood and alertness, and hue control.
- New and elaborated forms of light art and performance that can be shared and participated in locally generated, globally reaching culture where human to environment interactivity is displayed, intermediated with embedded intelligence, sensors, and AI in the cloud.
- Better dark sky control for urban stargazing and astronomy with more granular controlled distributed lighting.
- Individually more responsive luminaires and lighting provisionally controlled with cell phones of emergency workers, and even citizens as needed for safety, communication or fun within certain contexts.

The list of activities and scenarios affected or enabled by smart lighting is limited only by the imagination of lighting designers and creative practitioners.

The Current Reality

Cities around the world are exploring the new frontiers of Smart City Lighting with varying degrees of ambition and success.

If you look at the global night map (Fig. 11.3), you can see that many people on our planet would be happy just having electric light at night so their kids can study, extending work hours and providing basic safety from human and animal predators. Most underdeveloped countries have regular blackouts and limited power

Fig. 11.3 Lighting of the Earth [19]

distribution. The greatest need for light, besides the abrupt large shadow on the map of North Korea, is located in the tropics and subtropics where the days are always fairly short and there is an abundance of powerful sunlight. The promising technological leap-frog of solar and wind power, (as has already happened with the leap of cell phones over legacy telephone infrastructure), combined with the plummeting costs of LEDs and their DC compatibility with Solar power, is just starting to light up small cities and remote villages, and often includes cellular/WiFi transceivers in the development package.

New Smart City projects are being deployed all around the developed world. Successful projects include the UN partnerships with companies like Philips Lighting on the "Solar LED Lighting 1000 Villages Program." Other projects try to set up urban conditions for the "next Silicon Valley." These projects are happening worldwide and provide a high-bandwidth infrastructure with smart, high efficiency urban lighting. Though usually unlikely to exactly produce the hoped-for result because of the rather hard to reproduce conditions and culture of the San Francisco bay area, these pilot infrastructures are certainly providing great opportunities for the development and implementation of Smart City lighting and providing new business and development opportunities. This new infrastructure is leveraged for more control of interactive lighting and site sensor data. Lighting can also enable network densification by stationing WiFi and cellular nodes on street poles.

Lighting and IoT Pilot Programs in Cities

A number of IoT pilot programs in cities worldwide are developing novel lighting solutions as well as interoperability and business models. For example, the Digital Inclusion Fund attempts to provide broadband services to low-income residents, which leads to co-creation initiatives so citizens, businesses, and government can create the most relevant solutions. New partnership models are emerging where companies can innovate and share incubator spaces. In turn, software is being developed for better visualization so city stakeholders can more easily contribute to infrastructure planning.

Crowdsourcing approaches also provide entrepreneurs with the ability to bid directly to the citizens, and applications such as SeeClickFix provide descriptions of problems like dead streetlights and potholes which are automatically sent to the appropriate government agencies with geolocation technology. As a result, social media and email has become the most used channel for citizens to communicate to government. These new channels of communication are requiring a cultural shift in city government, both because it is 24/7 and the use of public funds and efficiency of government are becoming more transparent and subject to public scrutiny and criticism. The World Council on City Data (WCCD) was launched in 2015 to benchmark and compare the service and quality of life indicators in various cities.

Citizens are becoming more willing to share their personal data when it can be used to improve city services, transportation, and congestion. As a result, a new wave of services has been enabled by the launch of open data portals in many cities which creates additional opportunities for software developers and startups.

Companies Leading Projects in Smart City Lighting

Taking advantage of the movement toward Smart Cities and the ensuing opportunities in Smart Lighting is an important part of modern business. Several companies have launched platforms which leverage IoT and other technologies to provide smart lighting solutions. The aim of these platforms is to integrate appropriate technologies into viable, repeatable solutions for municipal customers. A few of the more notable platforms and projects are described in subsequent sections.

Acuity

The "ROAM system" has a mesh network with uplink via cellular wireless or Ethernet-connected nodes that replace the standard photocells and control and monitor each luminaire. They monitor voltage and current, and send that information to the gateway. Each gateway receives data and transmits commands to up to 2000 luminaire nodes. The Network Operation Center collects, stores, and analyzes the data for efficient control of the lighting network. It also safeguards and archives the data for future uses.

GE

The "Predix Platform" is GE's cloud-based approach to smart lighting. The platform is ambitious with the goal of integrating lighting control and sensor technology in their "Current" energy and data company. However, the Predix control solution is currently less developed, the idea being that other startups will develop new systems and apps around GE's basic infrastructure. This infrastructure includes LED streetlights, electric charging station with built-in solar panels, WiFi on the streets, various sensors of occupancy, humidity, and sound to generate meta data in order to mitigate safety incidents, track parking spaces and traffic, enable faster snow plowing, and improve energy efficiency. This is an open data system, so any city system can share and use it.

Philips Lighting

The "CityTouch" system currently has one of the most complete portfolio of LED luminaires, controls, software management platforms, apps, and services to support Smart City initiatives. This is a scalable solution for fast growing cities, with a future-proof IT strategy, and cloud solutions will be the next step. Philips Lighting also developed "SmartPole" that integrates 4G LTE cellular transceivers to streetlight poles and LED luminaires to enhance wireless broadband coverage to businesses and citizens, in multiple cities in the US. This improves mobile network performance and reduces urban clutter.

Sensity

"NetSense for Cities" is probably the most sophisticated city lighting/security control system. The concept is to deploy city-wide security with high efficiency LED lighting. This approach helps cover the cost through energy efficiency and lower maintenance costs due to the long life of LEDs and more precise maintenance notification. The security cameras and video analytics by embedded intelligence provide data to the NetSense integrated application to better control the luminaires for lighting optimization based on occupancy, as well as circuit monitoring, wire theft detection, smart parking for parking lot and street-wide applications, as well as general security.

Leading Cities with Smart City Lighting Projects

Leveraging novel solutions and integrating multiple technologies is the purview of future-looking cities worldwide. A number of cities have deployed systems and technologies which provide or enable smart lighting as part of their Smart City developments. The primary goal of these projects is to provide enhanced municipal services and increased visibility for the city's citizens and visitors. A few of the more notable projects are described in subsequent sections.

Barcelona, Spain

Some call it the world's smartest city. The industrial neighborhood of Poblenou is being revitalized to turn it into an innovation district that is knowledge-driven. The District is called 22@ and has a digital infrastructure and broadband and free WiFi that is attracting digital industries and tech-savvy residents. The same digital platform is being used in government on collaborative projects, which helps break down silos. Sensors installed in the urban infrastructure wirelessly provide data on temperature, air quality, parking meters, pedestrian traffic, and even waste bins.

Boston, Massachusetts

Boston has deployed a project called "Connected Bits" as a way for citizens to become more involved in the running of their city. A cell phone app enables citizens to easily spot and report graffiti, blocked drains and faulty streetlights, potholes, fly-tipping, (illegal dumping), graffiti, and antisocial behavior. This new tool of "Participatory Urbanism," allows the city to easily and quickly fix common urban problems.

Charlotte, North Carolina

The "Envision Charlotte" project is a collaborative effort between Duke Energy and Verizon, with Charlotte Center City. In this project, the partners are installing interactive kiosks and monitors to gather and feedback energy use information.

Eindhoven, Holland

Leveraging the vision for "Roadmap Urban Lighting Eindhoven 2030" and known as "The world's first crowdsourced Smart City," Eindhoven is pushing the envelope of Smart City projects. Eindhoven is where Philips is headquartered, and where they are continuing to do pilot projects of innovative city lighting, controls and IoT communications and sensing. Philips is trying out systems that guide emergency services to incidents, use sensors to more finely adapt lighting to weather conditions, and provide lighting on demand on the streets. They have just begun five pilot programs in different parts of the city, focusing on solutions that provide real value to citizens.

Fujisawa City, Japan

Fujisawa is known as the "Sustainable Smart Town (SST)". This municipal program involves Panasonic, the Fujisawa City local government and a number of business partners including Tokyo Gas, Accenture, and Yamato Transport. The partners are using IoT technologies to develop city services encompassing the important Smart City "pillars" of community, mobility, energy, security, and healthcare. Starting in 2014 in Fujisawa, 220 homes were designed from the ground up to enable these goals.

Kansas City, Missouri

In Kansas City, Cisco, Sensity, and Sprint are implementing a smart sensor and LED streetlighting system that dims automatically, saving energy and reducing light pollution, as well as free WiFi. The project is known as the "Smart City Corridor."

Los Angeles County, California, and Huntington Beach

The "Better Communities Alliance" in LA County is a way for cities and counties to access DOE and Ameri-Corps resources. More California cities are joining the alliance, including Chula Vista, San Francisco, and Sonoma County. The idea of the initiative is to deploy Smart City technologies in a responsible and equitable way. Using Philips "CityTouch" and connector nodes with plug and play activation, Los Angeles connected 110,000 light points.

Oslo, Norway, Dresden, and Klingenthal in Germany

The "City Tree" is an air-purifying, and digital display/WiFi/and e-bike charging station tower. This vertical, bioengineered, 4 m high stand purifies the air with power equal to 275 trees. The advantage over trees is definitely not aesthetic, however.

New York, New York

In the "LinkNYC" project, Google startup Sidewalk Labs is replacing old payphones with interactive kiosks. These smart kiosks provide WiFi, and services such as maps, emergency, and advice. Known as an "urban innovation company," Sidewalk Labs may also be an experiment in sociological phenomena. A few homeless people have also controversially used them for streamed pornography, gathering around the kiosks like a campfire in the snow.

San Diego, California

In San Diego, the city is revitalizing the Port of San Diego, the PETCO Park baseball stadium and other downtown areas via the installation of 75,000 intelligent streetlights, as well as solar to EV charging technology.

San Jose, California

Connected and web-enabled road lighting defines a main component of the San Jose smart lighting deployment. These lights are individually controlled and monitored for maintenance, and provide energy use, sound location technology (for gunshots), daylight controls, and occupancy dimming via a centralized control system.

Singapore, Indonesia

The Singapore "Smart Nation Initiative" started in 2014. Singapore has one of the highest mobile penetration rates. In this initiative, city systems use a common technical architecture to create a connection between people and city systems which leverages their mobile technologies. The system features an open data portal (data.gov.sg) which hosts datasets. The connected streetlight management system pilot is in the island resort of Sentosa, where 300 LED road luminaires will be installed over the next 5 years. The project expects up to a 50% improvement in energy efficiency, reduced downtime, and improved operational efficiency. These benefits are a direct result of remote monitoring and also produce better public safety with improved visibility and controlled light pollution.

Obstacles to Development of Smart City Lighting

According to Bill McShane, National Director of the Philips Connected City Experience for Philips Lighting, specific obstacles continue to slow the development of lighting technologies for the Smart City. Many of these obstacles are related to the lack of a clear understanding of the value-add of IoT and smart lighting for cities.

One major sticking point is a lack of consensus on the meaning of the term "Smart City." According to McShane, some believe a Smart City is "as simple as wireless electric meters, water meters, parking guidance systems, or LED lighting," while others see it as a "complex interconnected information gathering and surveillance system" including continuously captured video. Although video can be used for occupancy sensing for lighting as well as for surveillance, storage, and analysis can be problematic. The massive amount of data generated by large numbers of distributed sensors is also a cause for concern in terms of storage as well as privacy.

McShane comments further that city governments are often too conservative with the opportunity of new technology, relating that "city staff seem to move slowly and are resistant to change." This is understandable given the lack of a comprehensive, coherent public policy on data ownership and use. For example, can images of citizens, even license plates be sold, or does the identifying information need to be scrubbed? The reluctance of cities to adopt new technologies can also be traced back to widespread misinformation from smaller companies who have over-promised and under-delivered connected lighting to cities, leading to a climate of disappointment in the technology. However, although cities may be slow in adopting new methods of procurement required for the new scale and complexity of smart lighting, McShane believes that cities should be "open to new ideas, new ways of procurement methods such as requests for a bid or proposal, versus private public partnerships."

Cyber security and data volatility are also major issues, particularly when controls are connected via IoT technologies. The lack of a coherent public policy that supports IoT and can be implemented and managed by municipal personnel adds to the lack of market consensus on IoT technologies and standards. The result of this unsettled market state is extreme difficulty in providing data from sensors in a format that is useable for cities and that can ensure citizen privacy and security. The distributed nature of these systems leads to complexities in management and provisioning, particularly when considering "future proofing" and compatibility.

McShane muses on the number of "smart" systems: "smart grids, smart homes, smart environment, intelligent transportation, HVAC, security, communications?" The proliferation of intelligent, autonomous, and connected systems will be neither straightforward to implement nor simple to manage.

The Revolution Begins

We are in the pioneering stages of Smart City Lighting. There is great promise in improved efficiency, new, and better service, lower glare and increased safety. New possibilities are emerging which will improve the quality of urban life using intelligent lighting solutions.

For stakeholders, there is a legitimate security concern with Internet-based municipal systems which can be intensified by a lack of infrastructure funds in urban governments. Large technology suppliers are moving too slowly around such an obvious opportunity in this confluence of technology. Orchestrating the development and implementation of Smart City lighting requires breaking down long established silos, whether in industrial research and development labs, or in governmental departments.

As in all revolutions, Smart City lighting will continue to develop in fits and starts, with unexpected shortcomings, and new economic and quality of life opportunities emerging. Because we are in the middle of the revolution, we may hardly notice it happening around us in the developed world. In small tropical villages, the change will be much more dramatic and exciting, when the first solar power, LED lighting, and cellular/WiFi arrive in its most basic form.

The economics of lighting efficiency, the promise of the smart grid, the unmet need for connectivity, and the new form of gold ("data") will likely drive this revolution forward. Quicker than we realize, we will be in a much hotter pot of water, like the proverbial slowly cooking lobster. Technological development, though inherently democratizing in the long run, is morally neutral. The current and future use for the benefit of all is yet to unfold, and clearly we can create very different futures for ourselves with the same ingredients. Here are two imaginary scenarios.

Dystopia

We find ourselves living in an environmentally collapsing Big Brother/Dark Mirror nightmare with oppressive faceless overseers monitoring all citizens with infrared-lit street cameras everywhere, home televisions, home computers and cell phones, or even our intelligent, nearly sentient homes, observing and relaying with a variety of built-in sensors our every movement and occupancy pattern suspiciously, while pretending to only host and service us. We receive only heavily censored information and entertainment, as species die off and any future worth living is robbed from us and our progeny. Continuous and arbitrary war, the main economic activity and entertainment, maintains the lavish lifestyles of a tiny minority of arms merchants, national warlords, and media moguls. Humanity relentlessly poisons its own nest, or suddenly self-destructs in staccato flashes of light, most species going into oblivion with us. The globe's poorest and most remote populations mercifully remain in the dark, disconnected from our self-created nightmare as nuclear winter approaches

Utopia

We wake up to a globally interconnected well-lit, mostly urban, yet rural-inclusive democracy with healthy fresh food for all, grown with natural and artificial light in every neighborhood. The night view of the globe has more evenly distributed points of light over inhabited areas, with no permanent shadows hanging over populations of oppressed and impoverished millions. Countries compete in the World Happiness Index, instead of GDP, or arms race, aiming for the highest quality of life for citizens. Humanity's greatest natural resource of talent and creativity, interconnected through data and light co-creates a jewel of a humane, nature-loving civilization, finally worthy of our lovely planet.

Conclusion

The revolution in Smart City technologies is affecting every aspect of modern urban life. The incorporation of lighting solutions into active Smart City environments is just an example of the breadth and importance of this trend. From LEDs replacing conventional light sources and participating in mesh networks, to the unprecedented control of illumination via hue changes, to the efficiencies available from intelligent, responsive, large-scale lighting systems, the trend is expected to continue unabated for years to come.

This chapter has presented a vision for the future of lighting in the Smart City, as well as certain technology drivers and potential roadblocks. As the Smart City concepts and implementations progress, municipal lighting solutions will open up new modes of communication, wayfinding, and interaction at an urban scale, with controllable pixels, high-bandwidth communication, and endless possibilities for civic/artistic expression.

Acknowledgements Writing this chapter has turned out to be a summing up of my 11 year career in industrial design and product development with Philips Lighting, that ended happily, and with mutual gratitude in January this year. LEDs (light emitting diodes), were my first unreasonable new technology enthusiasm back in 2005, when they were overpriced, and still not nearly as efficient as High Intensity Discharge lamps. But they were digital, highly controllable, and Haitz law, (the equivalent of Moore's law for LEDs), projected a clear future where in a few years where they would take over the market, as they now have. Since product design and development in lighting used to take several years, back in the day, before CNCs and 3D printing were widely available, developing ideas way at the beginning of the wave was essential to have any chance of incorporating it effectively into a viable product. So I took on the challenging role, in what was then a very entrenched and conservative industry, of insisting we take a leading role in technological trends with prototypes, pilot projects and product, in order to create value for the customer and generate product and service differentiation, in an increasingly cluttered marketplace. In the last few years using IoT, controls, sensors, network densification and AI in embedded, networked and stand-alone lighting were to me clearly the path out of product and service mediocrity and commodification and into a valuable future. So researching, reaching out to experts in the field, and speculating, sometimes using more imagination than knowledge, about the next revolutionary wave in lighting, where we are now in the early part, is what this chapter is about.

References

1. Wikipedia. https://en.wikipedia.org/wiki/Haitz's_law. Accessed 25 Mar 2017
2. SXSW Interactive Festival (2017). https://www.sxsw.com/festivals/interactive. Accessed 20 Mar 2017
3. Shedding Light on LiFi. http://purelifi.com/news_media/lifi-white-papers. Accessed 31 Mar 2017
4. Welcome to the Smart City—Philips Lighting (2016). http://www.lighting.philips.com/main/systems/connected-lighting/connected-lighting-for-smart-cities.html. Accessed 30 Mar 2017
5. Wikipedia. https://en.wikipedia.org/wiki/Hanging_Gardens_of_Babylon. Accessed 25 Mar 2017

6. Brockman J (2011) Why cities keep growing, corporations and people always die, and life gets faster: a conversation with Geoffrey West. https://www.edge.org/conversation/geoffrey_west-why-cities-keep-growing-corporations-and-people-always-die-and-life-gets. Accessed 30 Mar 2017
7. U.S. Green Building Council (2017) Leadership in energy and environmental design. http://www.usgbc.org/leed. Accessed 29 Mar 2017
8. Bay S (2016) U.S. Funds solar panel pavement that generates power. http://www.sunbaypaper.com/national/environment/u-s-funds-solar-panel-pavement-that-generates-power . Accessed 12 Apr 2017
9. Solar Roadways (2015) New USDOT contract—November 2015. http://www.solarroadways.com. Accessed 12 Apr 2017
10. Hale T (2016) Luminescent bike paths light up town in Poland. http://www.iflscience.com/environment/luminescent-bike-paths-light-up-town-in-poland. Accessed 15 Apr 2017
11. Koerner B (2016) Lucept embedded lighting series. https://lucept.com/embedded-lighting-series. Accessed 15 Apr 2017
12. Lavinia (2009) The largest LED architecture project in the world: Yas Hotel in Abu Dhabi. http://freshome.com/2009/12/06/the-largest-led-architecture-project-in-the-world-yas-hotel-in-abu-dhabi. Accessed 15 Apr 2017
13. Vingiano A (2013) Scientists in China made glow-in-the-dark pigs. https://www.buzzfeed.com/alisonvingiano/scientists-in-china-made-glow-in-the-dark-pigs. Accessed 16 Apr 2017
14. The Naked Scientists (2013) Tattooed circuitry. https://www.thenakedscientists.com/forum/index.php?topic=49891.0. Accessed 16 Apr 2017
15. Niefer A (2015) Biohackers are implanting LED lights under their skin. https://motherboard.vice.com/en_us/article/biohackers-are-implanting-led-lights-under-their-skin. Accessed 16 Apr 2017
16. Ali Express (2016) Luminous clothing. https://www.aliexpress.com. Accessed 13 Apr 2017
17. Vincent Callebaut Architectures (2017) Paris Smart City 2050. http://vincent.callebaut.org. Accessed 14 Apr 2017
18. Painter K (2015) Urban fish farm expanding in St. Paul. http://www.startribune.com/urban-fish-farm-expanding-in-st-paul/292310561. Accessed 17 Apr 2017
19. Astronomy Picture of the Day (2000) Earth at night, November 27, 2000. https://apod.nasa.gov/apod/ap001127.html. Accessed 15 Apr 2017

Chapter 12
Smart Water Solutions for Smart Cities

Thomas Dickey

Introduction

Every article published today talks about achieving Smart City status. Competitions abound to bring major metropolitan areas and rural towns together on an equal playing field to implement best known practices. From bloggers to late night talk-show hosts, entrepreneurs to Fortune 50 multinational corporations are all talking about Smart Cities as if it were clearly defined as a crystal vase. The dialogue is truly energizing and there is no barrier to entry, one simply must have an idea or an opinion on how technology can be integrated into our daily lives to make our world a better place. Unfortunately, there's not enough storage space on our smart phones for all the apps required to make this dream come true … yet. Smart Water Networks in Smart Cities is an exciting intersection between technology and conservation where safety, efficiency and conservation will simultaneously benefit without the need for fundamental trade-offs.

Definitions and Drivers

There are several areas of development that form the framework for evolving Smart City strategies. The common challenge amongst Smart Cities is shared learning between cities, adapting best management practices and technologies from one city to the unique environments of another. There appears to be much less of the NIH ego associated with the Smart City theme and much more of the "we can make a difference if we implement that idea" approach. Mother Nature has added her own

T. Dickey (✉)
SensCo Solutions, Austin, TX 78759, USA
e-mail: tdickey@senscosolutions.com

S. McClellan et al. (eds.), *Smart Cities*, DOI 10.1007/978-3-319-59381-4_12

sense of urgency to the Smart Water dialogue as unprecedented drought situations have moved the conversation needle from convenient to urgent.

Transportation and Mobility

Transportation and mobility are key considerations in every city. Citizens must be able to be safely and sanely get to and from their place of work. Some cities have already tackled mass transit with rail, subway, and public transportation options, thereby establishing a framework on which they can explore ideas ranging from bike sharing to intelligent parking to green municipal cars. In contrast, there are growth cities which appear to focus on the Smart City agenda without first having the basic infrastructure building blocks in place. At times these challenges can easily co-exist without conflict or competition for limited resources, but at the end of the day, smart people must sit down and have serious conversations about real problems and real solutions and take action. The Smart City infrastructure for transportation and mobility will only be fully be realized when technology, psychology, and financial visionaries come together with a community leadership force that measures success based on people-impact accomplishments and not hours spent in focus groups and city hall committees. Transportation and mobility challenges are not limited to just people, energy resources such as water and electricity must also be transported from their source to the point of consumption.

Energy

Energy and everything that relates to energy management, from municipal efficiency to individual citizens monitoring their own usage have an important role in Smart Cities. Highly publicized debates over energy rights and entitlements are often distracting to the amazing progress that is being accomplished in these fields. Brown-outs and droughts are a reality that historically we have come to accept but innovative breakthroughs have reduced our tolerance for such conditions. Consider efficiency to be a key metric of how well a Smart City is utilizing its resources and now envision Energy on Demand. The right type of energy resource, be it electricity, safe water, petroleum products existing in the right format at the right location at precisely when required. Acquisition of new energy sources is wildly regulated, burdened with extreme timelines and in many cases, cost prohibitive. Consider being more efficient and less wasteful with the existing resources that are currently available. Smart Cities are putting the challenge of doing more with finite resources to the test and some of the results are brilliant.

Information and Communications Technology

Information and communications technology and digital disruption is the pillar where many cities have made their first moves. Cities which embrace and promote entrepreneurship are going to win if they can sustain the necessary financial and talent pools. In 1999, the trend was to take an idea (more bad than good), find a clever company name, add ".com" to the end of it and build a website. Raise enough money to otherwise provide humanitarian aid to thousands of needy people, rent a high-priced address and bring in the ping pong table. The stories of failure defined an era while a few heroes of commerce emerged. Less than 20 years later, we realized that those digital pioneers did not have the benefit of today's incubators, accelerators and co-working spaces. The explorers of today's "there's-an-app-for-that" era have huge advantages over the original digital pioneers and the results are spectacular. Not to say that there are not a few crazy ideas driven by the Smart City theme, but the technology platforms, tools and interoperability standards provide the foundation for real problems to be addressed with innovative solutions. It is through these technologies that active and meaningful participation is driven down to the individual citizens.

Humanity

Humanity rounds out the Smart City pillars and this is a game changer. Even the acronyms that are being recycled have taken on new meanings. BOGO was synonymous with Buy One Get One, but today it is widely recognized as Buy One Give One. TOMS© Shoes is known as the One For One company. Sky Footwear© is the Love Socks, Love People company. Both companies have architected business models that purposely put humanity into their Mission and Income Statements. Study the models of Home Away, Uber, Half Helen, and Roller Hippos and the themes of commerce and humanism are magically blended.

The concept of "Smart Water solutions for Smart Cities" embodies all four of these pillars and is considered one of the critical emerging themes which will ultimately define the success of every Smart City agenda.

Deployment Considerations

Smart Cities have many moving parts, some of which are tightly integrated and others of which are loosely connected.

Competition for limited resources tops the list of challenges faced by all Smart City related endeavors and many of these challenges do not have quick fixes. This competition is very healthy in many examples of Smart City commerce and highlights both the benefits and potential negative consequences of capitalism.

Innovation comes from all points along the commerce spectrum. Large corporations have the advantage of significant financial and human capital. These assets can overwhelm some challenges faced by Smart City agendas, but occasionally lack the agility to productize their intellectual capital. At the other end of the spectrum are the small entrepreneur teams that enjoy limited bureaucracy, rapid prototyping mentalities and laser focus but are handicapped by their ability to safely scale their solutions to a level that have material impact.

Enterprise level businesses deserve recognition for acknowledging their limitations and adapting their business models to compensate. The most visible area is the corporate venture capital (CVC) ecosystem which is expanding far beyond the likes of Google Ventures and Intel Capital. Companies in a range of industries are launching CVC arms, such as aviation companies like JetBlue, and healthcare companies such as Orbimed Advisors.

The number of new corporate venture capital firms making first-time investments reached record levels in 2016. 107 new corporate VC units globally made their first investment in 2016, including funds like Baidu Venture and Sony Innovation Fund. Since the beginning of 2010, over $150 billion has been invested in rounds involving CVCs, according to the PitchBook Platform (www.pitchbook.com).

There are now corporate venture capital funds that specifically focus on smart energy and smart water market segments. Funds such as Cleantech Ventures and Mercom Capital Group recognize the challenges faced by entrepreneurs who are in the early stages of commercializing new and innovative technologies into market segments traditionally dominated by large corporate enterprises.

The benefits to both the corporate ventures capital groups and entrepreneurs are important to recognize as a driving force behind both the local and national success of Smart City agendas. It takes a significant amount of investment to get a disruptive concept to market and scale, which makes sharing the risk and the upside among other investors attractive. By leveraging capital on the balance sheet through minority investments, CVC groups enable scaling of R&D efforts beyond what their operating budgets allow, at the same time, spreading the risk and capital exposure between multiple entities. CVCs will generally be easier to get in touch with than traditional venture capitalists for entrepreneurs as the CVCs do not typically see as many deals as professional venture capitalists.

As innovation continues to accelerate, companies that take an investor's view of the market create exposure to markets outside their core strategic vantage point. One key strategic advantage provided by the CVCs through their early stage investments is the insight to evaluate innovation both in core and adjacent markets. This serves as a feedback loop between emerging markets and corporate R&D teams. The entrepreneurs who find corporate partners beyond their target markets can often gain introductions and access to new segments and customers not achievable on their own.

When CVCs engage early with entrepreneurs, they open opportunities to collaborate on solving complex problems for existing customers in a more agile and focused approach. Changing market dynamics across multiple industries create new challenges best solved by smaller, more versatile companies, often with innovative solutions. In exchange, entrepreneurs can use CVCs as potential sales channels. This relationship can be a great opportunity to showcase the innovative approach to a potentially large customer or channel or customer.

Startups are often the source of disruptive innovation, which gives companies an opportunity to use their venture arm to tap into that energy and create brand alignment. This approach allows larger corporations to maintain their leadership status by fostering the ecosystem. CVCs can make great advisors, advocates, and mentors, and often increase credibility by sharing thought leadership platforms.

The most obvious benefit for both parties is a close tie creates a future acquisition opportunity. CVCs who are first to the funding discussion will often preclude other competitors from taking a stake, creating exclusivity. This is a natural timeframe for the due diligence process to take place in a more natural operational setting.

Established companies, big and small, should consider engaging in corporate venture as an engine of growth. The knowledge transfer and opportunity created by engaging with some of the most innovative players in the industry will undoubtedly benefit the company and allow a faster response to market transformations. In turn, entrepreneurs should leverage their place in the market to build a strategic relationship that can accelerate the growth of their startup.

Municipal Water Management

Austin, Texas covers 270 square miles, has the 11th largest population at just over 2 million citizens and is recognized by Forbes as the fastest growing American city. The city's 2016 water loss report shows more than 5.8 billion gallons of treated drinking water leaked from city pipes before reaching customers in 2015 [1, 2]. This represents a 23% increase over the previous year, a trend that needs to be reversed before it becomes a negative influence on the local quality of life.

To put 5.8 billion gallons into perspective, it is enough to fill Lady Bird Lake nearly two and a half times. Overall, the amount of drinking water lost from city pipes has nearly doubled since October of 2011.

The City of Austin uses about 128 million gallons of water every day. An average loss of 12%, translates into over 15 million gallons of lost water daily. These figures are for water usage and water loss up to the meter. In other words, water loss on the consumer side of the meter needs to be added to these figures to fully scope the impact water loss in the city. The water loss situation in Austin is summarized in Fig. 12.1, where the leakage in 2015 totaled 5.86 billion gallons. This is a staggering amount of loss, equivalent to 17,983 acre-feet or roughly 2.4 times the size of Lady Bird Lake.

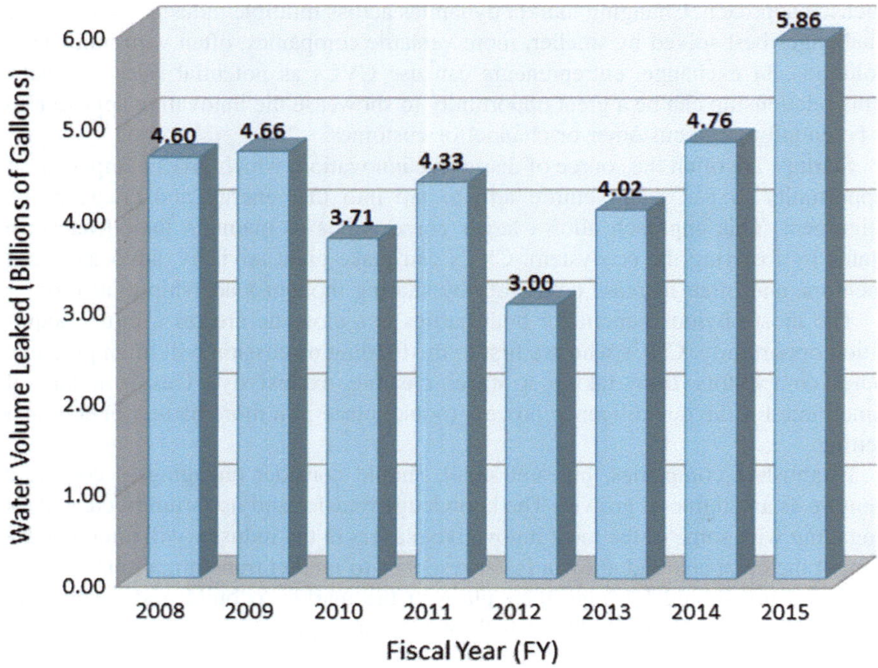

Fig. 12.1 Austin Water Leaks in 2015 [1]

Challenges

The challenges faced by the City of Austin are on both the supply side and the demand side of the water conversation. The Lower Colorado River Authority (LCRA) supplies Austin with surface water flowing from the Colorado River, managed through state granted water rights and a variety of incremental water supply contracts. The city manages two treatment facilities, capable of processing 285 million gallons of water daily. Water is delivered across a framework of nine major pressure zones, 35 pump stations, 34 reservoirs and a 3600-mile network of pipeline, supporting 25,000 public fire hydrants and 230,000 water meters. Additionally, the city keeps approximately 170 million gallons in storage.

The Austin population is growing just under 3% annually which strains the system in several ways. New homes being built in the suburbs require new infrastructure from pump to pipe to meter and then inside the home. The significant immigration destination is within the city core and this requires that the existing infrastructure capacity support an increase in demand.

Of the city's 3600 miles of pipeline, almost 750 miles are old cast iron pipes that are scheduled for replacement with new HPE sections. Current city budgets and staffing roadmaps estimate that the cast iron to HPE pipe conversion will take somewhere between 75 and 100 years at the current pace. The city's core is also under constant demolition of old structures and new construction to support the waves of new residents which puts an additional burden on managing the effectiveness of the water delivery network. Pipe leaks can also be difficult to detect. If water from a broken pipe comes to the surface, the break can be relatively easy to isolate. If a pipe is broken deep underground, and the water never reaches the surface, it can be much more difficult to find and repair.

Smart Cities Council, the world's largest smart cities network, recognized Austin as one of five U.S. cities to win its 2017 readiness challenge grant supporting innovation, inclusion, and investment in local communities.

Austin plans to design strategies and solutions for affordable housing for underserved populations, mobility, and economic development issues that have been heightened by Austin's rapid growth, booming tech sector and attractive quality of life. The next step is to transform from using technology in limited and specific ways to a broader and systemic approach to address community issues.

The immediate benefit to organizing these programs under a Smart City theme is the interoperability factor. Data collection, data transmission, data storage, and analytics as well as data delivery in terms of end-user platforms will require some broad level of compatible infrastructure. The challenge will become very complicated as competing technologies vie to become the standard of choice for the Smart City platform.

The number of sensors that will be collecting information will grow exponentially for the next several years. Consider the Google Driverless Car program that has over 650 sensors on board, collecting 1.3 million readings per second. Once those readings are collected and processed locally, some portion of that information will be transmitted into the cloud where it is analyzed and actionable commands are then sent back to the car where the message is redistributed and action is taken. Now let us take integrate the driverless car with the Austin City traffic network and Emergency Vehicle program to ensure the driverless car co-exists with the other 220,000 cars that travel in and through Austin on IH-35 daily.

A Sensor Network

What would happen if the 3600 miles of water distribution pipeline across Austin had sensors installed which were designed to monitor the flow of water and detect leaks? If 10 sensors were placed on every mile of pipe and internet gateways were placed on every fire-hydrant, there would be 36,000 sensors and 25,000 gateways installed across the city. The primary objective of this system would be reducing the mean-time to leak detection, reducing the mean-time to leak repair, and thereby reducing the volume of lost water.

If all new HPE pipe scheduled for installation (750 miles of pipe) already had sensors integrated into the pipe, no retrofit projects would be required for installation of sensors. If these sensors could transmit a signal up through 12 ft of caliche clay and limestone, concrete, rebar and asphalt and send data to the cloud, this data could be available for further processing and interpretation. These logistical challenges are layered on top of the assumption that the sensors that are spaced approximately 500 ft apart with frequent changes in elevation and direction can detect something that distinguishes a leak from flowing water from a train that is rumbling down the tracks directly above the buried sensor, and that reliable power is available to the sensors so they will not have to be dug up to have their batteries replaced.

All of this to reduce the time it takes to identify a leak and estimate the size of the leak so that it can be prioritized for repair with the other known leaks. The City of Austin recently implemented a Water Leaks Map [3] to share the status on known leaks from identification to repair. A screenshot from this system is shown in Fig. 12.2 where leaks repaired in the last 24 h, leak work-orders pending repair, and leaks pending inspection are updated in close-to-real-time.

SENSCO has created a solution which integrates best-in-class sensor, communication, artificial intelligence and drilling technologies to create a Smart City—Smart Water platform. The business model is best explained starting with the problem and moving to the solution. In Austin, water loss from leaking pipes is estimated at 5.8 billion gallons per year [1]. Residential water rates vary based on total consumption but are conservatively priced at $7.00 per 1000 gallons.

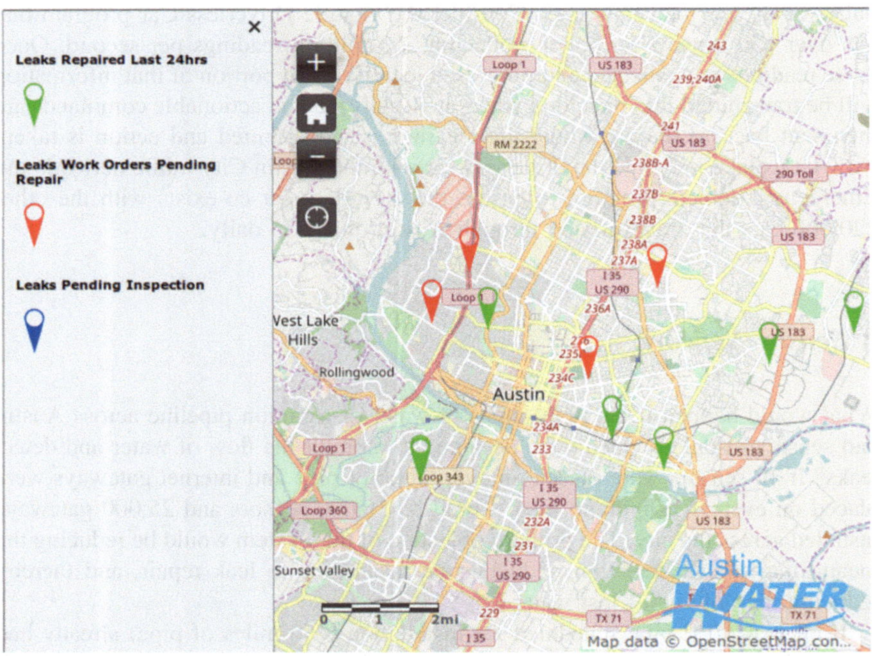

Fig. 12.2 City of Austin Water Leaks Map [3]

Recognizing the volume of water lost through a leak has two time components, the time from when the leak begins to the time that the leak is detected and then the time from detection until the time the leak is repaired. This is described in Eq. (12.1), where R_{leak} represents the rate of the leak, T_{detect} represents the time required to detect the leak, and T_{repair} represents the time required to repair the leak.

$$R_{leak} \times \left(T_{detect} + T_{repair}\right) = \text{Loss} \tag{12.1}$$

In many cases, the rate of the leak increases with time. In some cases, the rate can grow significantly if the leak turns into a burst. Intuitively, there is additional advantage to detecting a leak as soon as possible, although in this example we assume the leak rate remains constant.

If the T_{detect} is significantly greater than T_{repair}, then an early warning sensor system has great appeal compared with the scenario where T_{repair} is significantly greater than T_{detect}.

There is another scenario where the leak detection system would identify leaks that were not reported previously because the leak was either small enough of deep enough that it was never detected. In most reported cases, for small leaks the estimated T_{detect} is much longer than T_{repair}. However, for large leaks the opposite is true.

The Business Case

As a starting point for a business model, assume that T_{detect} and T_{repair} are the same and that the sensor network has 100% coverage and 100% accurate and immediate detection. As T_{detect} shrinks to zero, the water loss is cut in half, reducing the annual loss to 2.9 billion gallons.

The 2.9 billion gallons that is not lost in this scenario is not considered regained income. Rather it is recognized as 2.9 billion gallons of water that does not have to be processed. Operating costs to process and deliver a gallon of water to the home are estimated at $0.008 per gallon, making the 2.9-billion-gallon savings worth a whopping $23 million annually and this is just in Austin, Texas.

If the city were to accept a 5-year ROI , then the financials suggest that they could afford approximately $3200 per sensor as a fully burdened cost including installation and monitoring if the sensor network was 100% installed and fully operational on Day 1. If an aggressive 5-year deployment schedule were adopted, a realistic savings of $65 million over the 5 years would yield a $1800 per sensor installed cost.

Therein lies both the challenge and the opportunity. If both the municipality and the technology provider approach this problem with pure Income Statement mindsets, there is a significant risk that projects like this will never get past the discussion phase. However, if the cost of acquiring a new water source can be postponed through greater network efficiency and conservation, this opportunity begins to shine. If the cost savings realized by repairing minor leaks instead of major leaks is incorporated, the opportunity now shines bright. If the manpower previously deployed to find and repair leaks was retrained and redeployed to analyze the data being generated by the sensor network, the concept of moving from reactive repairs to proactive and preventative maintenance becomes an achievable reality. Smart Cities are going to force new and innovative business models between municipalities and technology providers in order to bring smart solutions into play.

Consumer Benefits

A similar model can be discussed on the consumer side of the water meter, where water is lost and wasted at a similar rate of municipality loss rates.

The City of Austin considers an efficient household of four with a sprinkler system to use 70,000 gallons of water per month [4]. Efficient households are those considered in the lowest 20% of monthly water users. Clearly the allocation changes on a case by case basis, but the typical residence (home owner, apartment or condo tenant, and trailer park residence) in Austin uses 25–40% of their total water for personal and cooking purposes and 60–75% for irrigation. This holds true for homeowners and apartment dwellers alike as the lawns and pools are the major categories for irrigation usage.

Making the residential model even more interesting is the fact that residential insurance claims rank flood damage from broken pipes as the number 2 expense category behind wind and hail damage claims. Hail, wind, and plumbing or appliance leaks followed fire as the most expensive claims [5].

Small- and medium-sized businesses, commercial buildings, and campus complexes (hospitals and schools) all have water consumption and loss models that warrant leak detection and conservation strategies.

Smart Cities have many citizen categories ranging from individual residence to large corporate enterprises. The needs and priorities vary significantly so the processes and forums in which these issues are addressed and the leadership that prevails over the Smart City ecosystem are most critical to the overall success. Smart Water in Smart Cities is a fundamental requirement as the urban population growth is going to quickly overtake the capacity of the infrastructure required to support it. While new water sources are an option, droughts quickly disrupt that approach and create crisis level responses. Changing the way that water loss and water waste are addressed should be the first line of action for Smart Cities. Addressing the challenge with an understanding that some components of the business model may not be instantly quantifiable, the governing leadership must prepare for the future by investing in the infrastructure today.

Conclusion

Smart Cities have opportunities to create pilot validation ecosystems for technologies being discovered in universities and commercialized through startup enterprises. Product validation, capital funding, and resource management for scaling are inherent in the entrepreneur's world but Smart Cities can create the framework under which these resources are available and deployable for the greater good of the community.

This chapter has presented and discussed the concept of Smart Water Solutions for Smart Cities. This concept embodies all of the central themes of evolving Smart City strategies, and provides a framework for future expansion of technology-enhanced infrastructure solutions. In addition to motivating the business priorities for smart water solutions, the chapter makes the case that effective water management is a critical consideration in the success of every Smart City agenda.

References

1. Barer D (2016) Austin water pipe leaks could fill Lady Bird Lake twice, in 2015. http://kxan.com/2016/05/16/austin-pipe-leaks-could-fill-lady-bird-lake-twice-in-2015. Accessed 30 Mar 2017
2. Texas Living Waters Project (2016) Texas water conservation scorecard. http://www.texaswaterconservationscorecard.org/. Accessed 30 Mar 2017
3. Austin Water (2017) Austin Water Leaks Map. https://austintexas.gov/waterleak. Accessed 30 Mar 2017
4. Texas Living Waters Project (2016) Simple water conservation tips. https://www.nwf.org/~/media/PDFs/…/TLW_simplewaterconservationtips04.ashx. Accessed 2 Apr 2017
5. Insurance Journal (2016) Top homeowners insurance claims. http://www.insurancejournal.com/news/national/2016/04/06/404400.htm. Accessed 2 Apr 2017

Chapter 13
Technology-Enhanced Infrastructure

Peter John Schemmel, John Joseph Schemmel and Evan Diane Humphries

Introduction

The backbone of every city is its system of public and private infrastructure assets located both above and below ground. Common examples, as seen in Fig. 13.1, include transportation facilities, academic campuses, office and commercial developments, hospitals and health centers, utilities, water and waste treatment plants, and storm water management systems. These assets allow for the mobility, education, employment, and safety of a city's population, which in turn impacts quality of life. Therefore, it is incumbent upon a city's leadership to maintain and improve the safety, functionality, environmental impact, and economic value of its infrastructure assets. Today, this can be accomplished by deploying asset monitoring technologies and implementing associated management strategies to ensure the efficiency, effectiveness, and accessibility of these assets. When properly maintained, the built infrastructure can then support applications such as a smart mass transit networks, flood control systems, or electric power grids.

In broad terms, technology- enhanced infrastructure (TEI) systems consist of four key components as outlined in Table 13.1. Figure 13.2 depicts the interrelationship between components. The underlying objective of implementing such a system is to improve the resilience of each structural element within the city's

P.J. Schemmel
School of Engineering and Physical Sciences, Institute of Photonics and Quantum Sciences, Heriot-Watt University, Edinburgh EH14 4AS, Scotland, UK
e-mail: p.schemmel@hw.ac.uk

J.J. Schemmel (✉) · E.D. Humphries
Ingram School of Engineering, College of Science and Engineering, Texas State University, San Marcos, TX 78666, USA
e-mail: john.schemmel@txstate.edu

S. McClellan et al. (eds.), *Smart Cities*, DOI 10.1007/978-3-319-59381-4_13

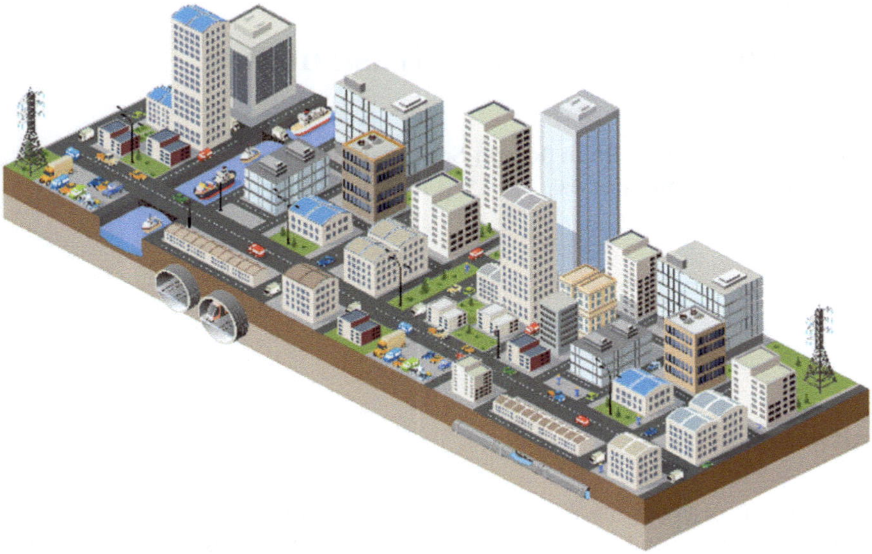

Fig. 13.1 Smart applications layered on technology-enhanced infrastructure

Table 13.1 Components of a technology-enhanced infrastructure system (TEI)

Component	Examples
Array of physical assets	Tunnels, pipes, foundations, bridges, buildings, industrial plants, power grids
Inventory of embedded, attached, and remote sensors	Strain gages, fiber-optic sensors, piezoelectric sensors, eddy current sensors, MEMS sensors, video cameras
Robust communication and data storage network	Wired and wireless communication, internet of things, data centers
Cohort of asset managers	Owners, civil engineers, environmental engineers, facilities management engineers, state, and municipal officials

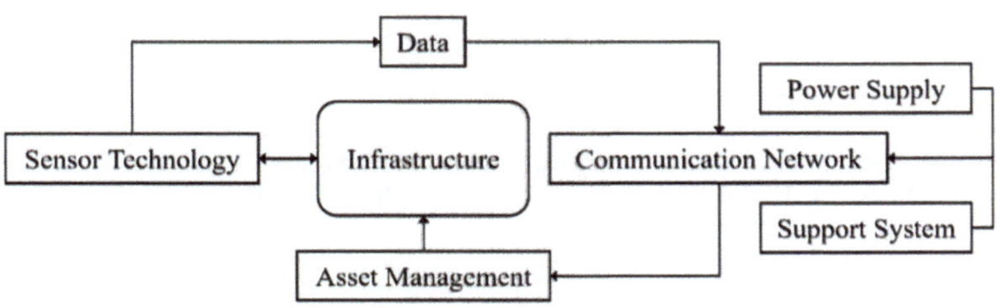

Fig. 13.2 Technology-enhanced infrastructure component interaction

domain. In so doing, this will enhance safety, prolong the useful life, and increase the value of an individual infrastructure asset.

In a functioning system, physical or environmental forces will activate sensors associated with an asset. In turn, these sensors will transmit measured data over a communication network to an asset manager. That data can be displayed and analyzed to make real time, well informed, decisions regarding the maintenance of the asset. This scenario can take place during construction, daily usage, continued maintenance, or throughout the entire life of an asset. For example, stress monitoring could detect an impending overload on a bridge that is under construction, vibration sensors could initiate dampers on a tall building foundation during high wind events, and repairs could be initiated when material deterioration of an underground pipe reaches a predefined limit. In each case, safety, functionality, and asset value would be maintained or improved as a direct consequence of technology being merged with the necessary structures within a city. Thus, the overarching benefit of developing a TEI system is that engineers, city officials, and the public can utilize the components outlined in Table 13.1 to better design, construct, manage, and maintain a city's assets.

Full benefit of TEI will be realized when it is overlaid with other technologies. For example, Adibhatla, Henke, and Atwater of ARGO Labs are developing a Street Quality Identification Device (SQUID) [1]. They have designed a sensor platform that is mounted on city vehicles and measures the surface quality of the roadway beneath the vehicle. Included in their system is a microcomputer, accelerometer, GPS, and camera. In another example, the 2017 Ford Fusion V6 Sport will come with a pothole mitigation system where sensors will monitor road conditions and activate suspension dampers when a pothole is detected [2]. In both cases, a remote sensor works with other technologies to capture real time information on roadway conditions. If desired, the data collected by such vehicles could be transmitted electronically to the city and stored in a database. This information could then be used by city officials, and other constituents, to make strategic and operational decisions regarding repairs. Technologies such as these improve safety as lane closures for manual inspections are minimized, enhance functionality as asset managers are better able to prioritize and optimize repair efforts, and provide environmental and economic advantages as the consumption of virgin natural resources and emission of CO_2 gases are reduced, service life is extended, and lifecycle cost are reduced.

While the benefits of TEI are clear, constructing a system has many challenges. Sensors are designed to be embedded within an asset during construction, attached to any asset once constructed, or located at some distance removed from the asset. Sensors can be embedded in concrete during new construction or repairs, placed within soil prior to compaction, or located within a body of water. However, most cities have extensive built infrastructure already in place, so easily attached and remote sensing devices are particularly needed for a wide range of applications. Table 13.2 presents material, physical, and environmental characteristics that might be associated with an infrastructure asset which are measurable with commercially available sensor technologies. The information presented Table 13.2 is based on

Table 13.2 Commercial sensors applicable to civil engineering assets (legend below)

Measured characteristics	Environmental	Structural	Transportation	Water resources	Geotechnical
Temperature	7 or more companies	7 or more companies	7 or more companies	1 to 3	1 to 3
Pressure	7 or more companies	4 to 6	4 to 6	4 to 6	1 to 3
Humidity	7 or more companies	4 to 6	4 to 6	None	None
Level/Elevation	4 to 6	1 to 3	1 to 3	4 to 6	None
Flow	4 to 6	1 to 3	1 to 3	4 to 6	1 to 3
Light/Visibility	4 to 6	None	1 to 3	None	None
Position/Motion	1 to 3	1 to 3	1 to 3	None	1 to 3
Sound/Vibration	1 to 3	1 to 3	None	None	1 to 3
Power	1 to 3	1 to 3	1 to 3	1 to 3	None
Water Presence	4 to 6	1 to 3	4 to 6	None	None
Gas Presence	1 to 3	None	None	None	None

Legend	
	7 or more companies
	4 to 6
	1 to 3
	None

data gathered from 27 companies that manufacture at least one monitoring sensor usable with civil engineering infrastructure assets.

Critical to the development and use of sensors is the identification of desired performance data for each individual infrastructure element. Sensors must provide data which asset managers want, can interpret, and that impact asset performance. Further, both external and remote sensors will need to be environmentally resistant and secure from damage and tampering.

It can be seen from Table 13.2 that gaps exist in the availability of some desired technologies. Thus, scientists and engineers will be relied upon to design and power cost effective sensors that in many instances do not yet exist. All the sensors within a city will collectively be transmitting several terabytes, or more, of information to asset managers in real time. This flow of information will only escalate as smart phones, vehicles, and other yet unrealized systems continue to merge with a city's enhanced infrastructure. Communication networks will need to have the capacity to transfer vast amounts of data from numerous infrastructure locations to data storage centers. Powering the large numbers of sensors and their communication networks will require new resources such as solar, tidal, and vibrational energy harvesting systems currently in development. With the potential for asset managers to be challenged by the volume of data transmitted to the city, it will be an absolute necessity to reduce all data streams into meaningful quantities and parameters so they do not become overwhelmed. Managers will also be challenged to interpret a sensor's data and subsequently determine an appropriate response based on that information. For example, does a signal from a limit detector suggest a serious safety risk or is it merely identifying an expected decline in structural performance? Even with all of these challenges, the ultimate goal is to have systems capable of monitoring any desired function within a city, which provide information to experts, who in turn make operational decisions based on public safety, lifecycle cost, asset functionality, and asset value.

This chapter examines the components of a TEI system, less the design or construction of the infrastructure asset. Sensor technology will be discussed first, including examples such as fiber-optic strain and temperature sensors, thermographic and radio imaging, and self-monitoring materials. Communication network issues such as support systems and power management procedures will be examined next. A review of asset management and data interpretation follows. Examples of functioning TEI systems are then presented. As indicated by Table 13.2, there are numerous sensing systems capable of monitoring a wide range of properties and characteristics across the breadth of civil engineering related infrastructure assets. This chapter is not intended to be a comprehensive examination of a complete TEI system. Rather, the chapter takes a snapshot of each component of a modern TEI system and provides insights into each.

Sensor Technology

The technology enhancement of an infrastructure asset begins with a monitoring device, or sensor. Devices can be embedded within an asset, attached externally to the asset, or placed remote to the asset. These devices respond to physical, chemical, and biological changes in an asset, convert those changes into an electronic signal, which is then emitted for detection. Some sensor systems are already familiar such as accelerometers in cellular telephones which are capable of measuring vibrations, anemometers that measure wind speed, or Global Positioning Systems (GPS). While these systems have matured, and are functionally reliable, they monitor and sense the environment surrounding an infrastructure asset. Acknowledging that this is important information, modern senor technologies need to focus on detecting, measuring, and transmitting data related to the physical properties and performance of an infrastructure element.

As suggested by Table 13.2, there several modern sensor systems applicable to a civil engineering infrastructure asset. Fiber optic Bragg gratings (FBGs) are an example of an embedded sensor. Radio antenna and thermographic imaging systems are a type of sensor that is positioned remotely to the asset. In some instances, such as self-monitoring concrete, the asset itself can serve as a sensor. Every sensor has its own manufacturing and operational benefits and challenges. Sensors that are embedded must be rugged and durable to transmit reliable data over an intended period during the life of an asset. While remote systems may not need to be as rugged, they are often more complex, are subject to variable environmental conditions, and require an operator with the associated potential for human error. There are also several categories of information measurable by modern sensor systems. Devices such as FBGs can be used to measure strain or temperature within a material. Radio or thermographic imaging systems can be used to detect cracking and delamination in materials. Still other devices can measure corrosion and salinity levels. Selecting the right sensor for an application will be critical to the usefulness of the data delivered to the asset management team.

Fiber Bragg Gratings

Since the late 1980s significant advances have been made in the development of FBG devices. The wealth and maturity of research on FBG systems makes them well suited for inclusion in TEI systems. Consequently, their use in infrastructure applications continues to expand. As shown in Fig. 13.3, FBGs are optical fibers containing a section with periodic variations of refractive index in the fiber core, typically written by a two-beam UV interference pattern [3].

Figure 13.4 depicts the grating structure which reflects light of a certain wavelength, known as the Bragg wavelength. The Bragg wavelength is dependent on the period of the granting structure as seen in Eq. 13.1 [4, 5].

$$\lambda_B = 2n_{eff}\Lambda \tag{13.1}$$

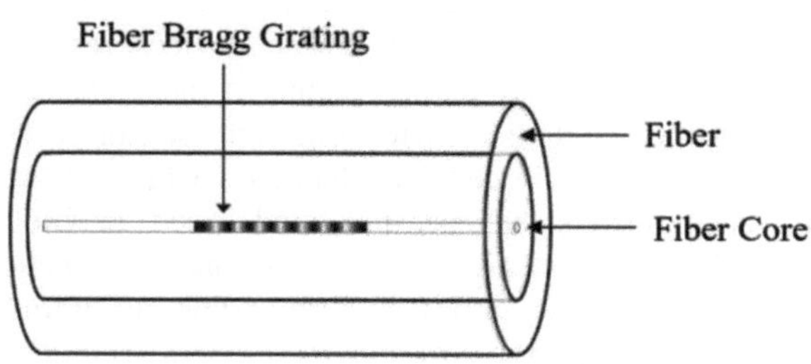

Fig. 13.3 Fiber-optic Bragg gratings

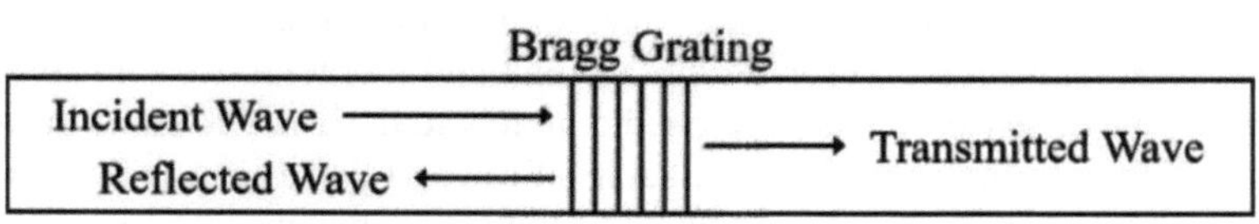

Fig. 13.4 Reflections from a Bragg grating within a fiber-optic cable

where λ_B is the Bragg wavelength, n_{eff} is the effective refractive index, and Λ is the grating period.

Variations in the Bragg wavelength can be used to detect changes in strain or temperature which alter the FBG period. Changes in the Bragg wavelength due to strain and temperature may be written as [4, 5],

$$\frac{\delta\lambda_B}{\lambda_B} = (\alpha + \zeta)\Delta T + (1 - \rho_e)\varepsilon \tag{13.2}$$

where α is the coefficient of thermal expansion, ζ is the thermos-optic coefficient, ΔT is the change in temperature, ρ_e is the effective photo-elastic constant of the optical fiber, and ε is the strain.

FBGs have several advantages for structural health monitoring applications. These devices are small and easily embedded, or mounted to, construction materials such as concrete, timber, metal, or carbon fiber reinforced polymers. FBGs are self-referencing devices and are less sensitive to power fluctuations from their control system. Further, they require less equipment to operate as only measurements of the reflected light are required, thus making a separate detector system unnecessary. Moreover, power and data acquisition wires are not required as the optical fiber acts as both the sensor and data communication line. Finally, FBGs are robust and offer excellent measurement accuracy. Monitoring strain in pavements, bridges, buildings, tunnels, and dams are common applications for FBGs. Several projects using FBGs to monitor bridge structures have been reported in the literature [5–7].

Although FBGs are primarily used for structural health monitoring they can also be used during construction and for in situ measurements of concrete shrinkage [4]. FBGs can be attached to reinforcing bars in structural piles, for example, prior to the placement of concrete [8]. Later, strain measurements can be recorded along the length of the pile over some time interval. Modified FBG systems may be used to detect moisture and humidity [4, 8]. While FBGs are not normally sensitive to water ingress, they may be coated with chemicals that swell in the presence of moisture. As the coating expands it changes the FBG dimensions which in turn alter the grating period, thereby measuring moisture content. FBGs placed in fresh concrete have also been used to monitor shrinkage, something not possible with standard strain gages. Finally, FBGs have also been applied to the measurement of geodynamics such as ground stress and strain, seismic activity, and road surface temperatures [4].

Supplying power and communicating with FBGs is very straight forward as power is required for only one central control hub to send light, and measured data, across each fiber. One major drawback to FBGs is that they require a large amount of wiring for an installation. This increases costs primarily due to the labor associated with installing a system. Regardless, it is expected that FBGs will continue to be a mainstay in TEI systems. Moreover, there is potential for FBG's to become the most widely used sensor technology for these kinds of applications.

Self-monitoring Concrete

Self-monitoring materials are those which can be used to detect changes in internal strain or damage without the need for a sensing device [9]. Carbon fiber reinforced cement, mortar and concrete are examples of self-monitoring materials that are of increasing interest over the past three decades. In these materials, randomly orientated carbon fibers, typically around 5 mm in length, are distributed throughout the bulk cement paste with the help of methylcellulose and silica fume [10]. Because the cement paste has a lower conductance compared to the embedded carbon fibers, small amounts of tensile strain initiate slight fiber pull-out resulting in an increased volume resistivity. On the other hand, large-scale damage in the form of cracks increases the chance that carbon fibers will contact each other, decreasing the volume resistivity. And finally, failure of the structure induces fiber breakage, again increasing the volume resistivity [11].

Measurements are typically carried out using the four-point technique where the two outer contacts supply a DC or AC current, and the two inner contacts measure voltage. Because the current and voltage are known, Ohm's law may be used to determine the resistance.

Self-monitoring materials provide a unique sensing technology for TEI systems. Not only does the material facilitate simple structural health monitoring, it also has high flexural strength and toughness, low shrinkage, and improved durability [12]. Furthermore, numerous measurements can be made quickly without the need to relocate equipment. The voltage measuring devices are compact, lightweight, and inexpensive. However, supplying power to the DC or AC current is disadvantageous.

Research shows that self-monitoring concrete is well equipped to provide valuable data to TEI systems such as for traffic monitoring and weighing in motion [13]. Self-monitoring concrete, in conjunction with digital imaging for license plate recognition, can replace weight stations for semi-trucks. The ability to weigh trucks in motion is more efficient than diverting them to a dedicated weigh station. Furthermore, self-monitored weight points can be managed and controlled remotely by just a few individuals.

Radio Wave Measurements

Radio and microwaves, roughly 30 MHz–300 GHz (Fig. 13.5), are often used in nondestructive testing applications due to their nonionizing nature and ability to easily penetrate nonpolar, nonconducting, dielectric materials such as wood, concrete, and engineered plastics [14]. Dielectric materials are characterized by their complex permittivity or refractive index [15]. When an electromagnetic wave is incident on a dielectric material, a portion of the wave is reflected at the initial boundary and the remaining wave propagates through the material as shown in

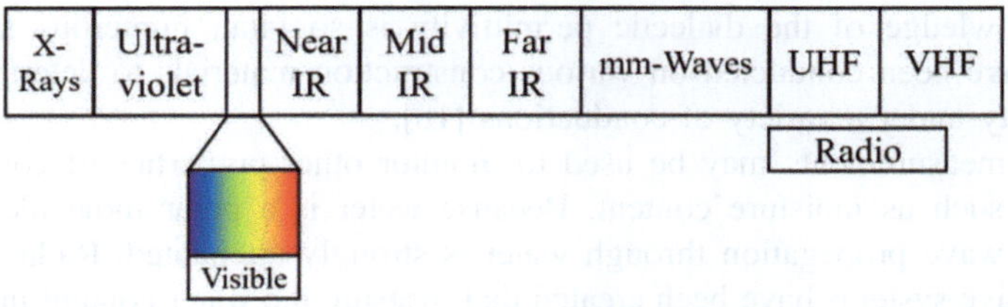

Fig. 13.5 FBGs work in the visible regime of the electromagnetic spectrum. X-ray wavelengths are approximately 10 pm, FBG wavelengths are on the order of several hundred nanometers, while radiation in the VHF region will have a wavelength around 10 m

Fig. 13.6 Transmission and reflection of an incident electromagnetic wave from a dielectric material

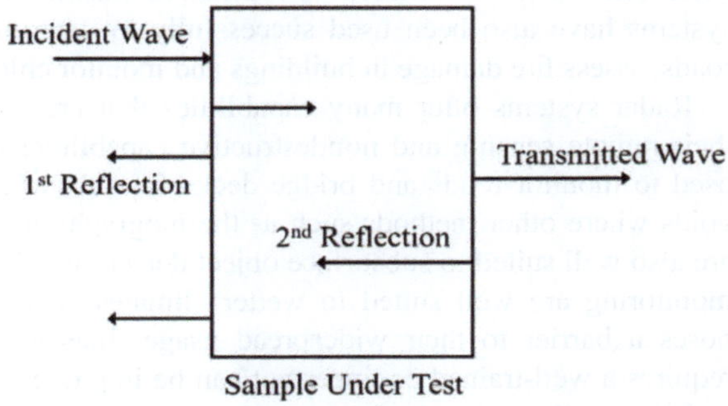

Fig. 13.6 [16]. If the material is semi-conductive transmission losses will occur. Measuring reflected or transmitted power as a function of time or frequency provides valuable information to inspectors.

Radar techniques have become common place in civil engineering applications such as bridge deck monitoring or for locating steel reinforcement [17]. When monitoring bridge decks, subsurface cracks or voids in the concrete create additional dielectric boundaries that reflect incident radiation. The same is true for locating reinforcement, where embedded steel bars act as small mirrors reflecting a portion of the incident radiation. Ground penetrating radar (GPR) systems, operating between 500 MHz and 3 GHz, are often used to conduct such tests. These systems range from simple hand held antennas to complex large-scale arrays towed behind a vehicle. Signal-processing techniques are typically required to present the measured data in a meaningful way. In addition, an understanding of the surface's dielectric properties is needed in order to make quantitative measurements.

GPR systems may consist of separate transmitting and receiving antenna, or a single transceiver, in the form of close contact dipoles or horn antennas. Measurements are made by emitting a wave from the transmitting antenna into the material. Receiving antennas are used to measure any reflected power signals as a function of time or frequency. The time it takes for a signal to be measured by the receiving antenna depends on the depth of the dielectric boundary and the wave propagation velocity determined by the dielectric permittivity of the bulk material.

Since knowledge of the dialectic permittivity is so vital, numerous laboratory studies have been conducted on various construction materials to determine their permittivity under a variety of conductions [18].

Radar measurements may be used to monitor other properties of construction materials such as moisture content. Because water is a polar molecule, electromagnetic wave propagation through water is strongly attenuated. Radar and millimeter wave systems have been created that measure the water content in concrete and timber for example. Furthermore, higher frequency millimeter wave systems are capable of imaging water ingress [19]. As a corollary, concrete curing can also be monitored by radar systems. Because concrete strength is inversely proportional to the amount of unbounded free water in the material, radar systems can be used to determine when the concrete has reached its maximum compressive strength. Radar systems have also been used successfully to determine the surface roughness of roads, assess fire damage in buildings and monitor chlorine content in concrete [16].

Radar systems offer many capabilities that are useful to TEI systems such as their remote sensing and nondestructive capabilities [20]. Such systems could be used to monitor roads and bridge decks for subsurface cracks, delaminations and voids where other methods such as thermography are not possible. Radar systems are also well suited to subsurface object detection, while water and chlorine content monitoring are well suited to wetter climates. The complexity of these systems poses a barrier to their widespread usage. Interpretation of measured data also requires a well-trained engineer but can be improved with robust signal-processing techniques. Data transfer from antenna systems is also straight forward in the form of raw data or processed images.

Thermal Techniques

Thermography is a common technique used to measure infrared radiation patterns from objects, with wavelengths around twenty times longer than visible light, typically between 3 and 14 μm [21, 22]. See Fig. 13.7.

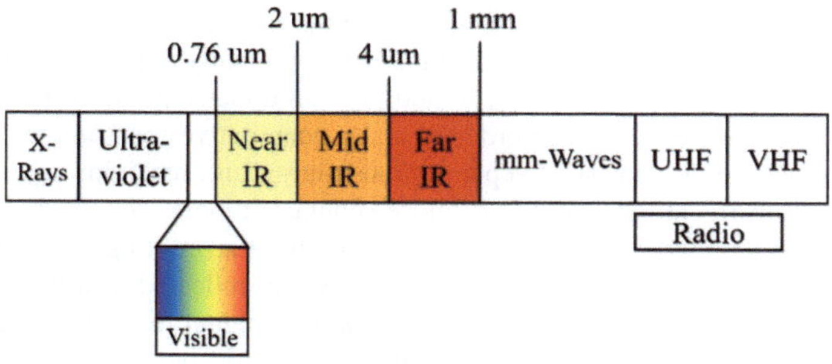

Fig. 13.7 Thermography works in the infrared regime of the electromagnetic spectrum

All objects with a temperature greater than absolute zero emit infrared energy, with a radiation flux per unit area of [23],

$$E = \varepsilon \sigma T^4 \tag{13.3}$$

where ε is the emissivity of the material, σ is the Stefan–Boltzmann constant, and T is the temperature. Comparison of measured temperatures of different materials must be conducted with care due to possible differences in emissivity, or the material's ability to radiate heat energy [23].

Both passive and active thermographic imaging techniques exist [22, 24]. Passive techniques do not rely on external heat sources and can detect both hot and cold sources embedded inside a material resulting in an increase, or decrease, in surface temperature. An example is the restive heating generated by electrical wiring inside a wall. Furthermore, established and stable thermal gradients, such as the warmer inner and colder outer walls of a heated house in cold climates, may be used to detect variations in thermal conductivity with passive systems. From a civil engineering perspective, changes in thermal conductivity, or a measure of a material's ability to transfer heat, are often caused by voids in a material [25]. Since air acts as an insulator, resisting heat transfer, voids in a material will cause a measurable change in surface temperature.

Active thermographic systems are useful when measuring objects in thermal equilibrium while incorporating an external heat source such as halogen lamps or even the Sun [23]. Active systems typically heat a material for a specific time interval and then use thermal cameras to measure the rate of cooling. Variations in the temperature response over the cooling period can be used to detect defects such as cracks, delaminations, and voids. More complex active pulsed systems capable of providing better quantitative data such as defect size, depth and location have been researched [25].

Thermographic systems are well suited for inclusion in most TEI systems [26]. These systems are relatively inexpensive, they can be placed several kilometers away from the object under inspection, and can provide accurate data easily transmitted across networks as image files. Thermographic systems do suffer from some disadvantages, thus making them ill-suited for certain applications. Environmental conditions can have a significant impact on the data collected from a thermographic system. Humid climates attenuate radiated thermal energy more than their dry arid counterparts. Standing water on a bridge deck will interfere with capturing thermographic data and may mask structural faults. Further, high winds can cause temperature shear effects, thereby complicating the data analysis. Lastly, a trained individual is required to interpret thermographic data.

Embedded Electrical Sensors

The durability of construction materials, especially concrete, is adversely affected by corrosion [27]. A major contributor to corrosion within concrete is the ingress of water contaminated with chloride. This water presents a threat to the steel reinforcement as it will likely cause corrosion. If the reinforcement in fact corrodes, this condition often causes other cracking, spalling, and delamination of concrete [28]. One method for monitoring moisture ingresses, and any corresponding corrosion of the reinforcement, is with electrical conductivity measurements [29]. While aggregate particles in concrete are nonconductive, cement paste is ionically conducting. Measuring concrete conductivity, and monitoring changes over time, can help determine the amount of corrosion, the corrosion rate, the extent of macroscopic cracking, temperature, and water penetration depth. As shown in Fig. 13.8, these measurements are typically obtained using embedded electrode array sensors which consist of a series of electrodes surrounded by a nonconducting enclosure such as Perspex or hardened epoxy resin [28].

Each sensor contains one or more inert electrodes and a reference electrode. Before deployment the electrode array is calibrated by measuring its response in various solutions with a known pH. Once calibrated the sensor can be installed during construction, or it may be placed in a core sample hole and secured with porous cement.

Fig. 13.8 Embedded corrosion instrument ECI-2 (Courtesy of Virginia Technologies, Inc., http://corrosioninstrument.com/eci-product-design/)

Embedded electrode array sensors provide a unique monitoring option for TEI systems. These sensors allow asset managers to evaluate the interior of a concrete element and repair that element before significant damage has occurred. This type of sensor system is akin to an early warning radar system that informs asset managers of impending degradation areas. If caught early enough, it may be possible to reverse the corrosion process with chemical treatments. These sensors have the disadvantage of not being able to be easily repaired should they malfunction. This should be taken into consideration when designing a TEI monitoring system. Finally, embedded electrode array sensors system have been utilized in conjunction with remote monitoring internet based systems, allowing users to monitor data from structures several hundred miles away. Thus, a TEI system will require a properly configured data communication network, professionals to interpret the data, and managers to maintain an asset for the TEI system to be effective.

Communications, Power, and Asset Management

As TEI systems mature, greater emphasis is being placed on the development of independent sensors packages. These systems incorporate self-contained power supplies, power storage, microprocessors for data processing, and wireless communication capabilities to transmit data. These issues have led to more extensive research into topics such as energy harvesting, power supplies, and wireless networks. Not lost in the focus on technology is the need for decision makers to use the data collected to maintain a city's inventory of infrastructure assets.

Communication Networks

An integral part of a TEI system is the ability to transfer measured data from the sensor to the user. Traditionally, this was accomplished by hardwiring sensors to a data acquisition system which was connected either to a single computer or possibly to the Internet [30]. Data was then transferred from the sensor and ultimately stored onto remote location servers accessible by users via a web-based protocol. The embedded electrode array sensor system, reviewed above, is an example of this kind of communication network. However, as TEI systems develop there is an understanding that the associated communication network needs to be wireless from the sensor directly to a data storage location [31]. This provides several advantages. A primary advantage of wireless communication systems is the ability to locate sensors anywhere in or on a structure. Second, wired connections become less feasible as the number of sensors increase. Cost, maintenance time, and the chance for malfunctions all increase along with the number of wired sensors. Despite the clear advantages of wireless communication networks, there are several challenges in developing a system that is robust and reliable enough to relay large volumes of precise data [32].

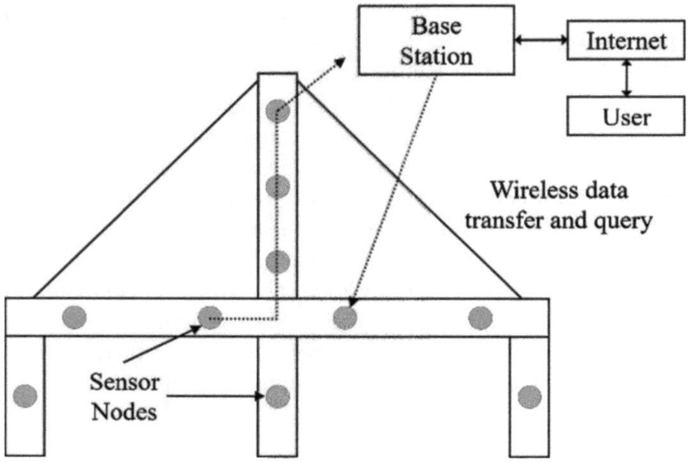

Fig. 13.9 Functionality of a wireless sensor network associated with an infrastructure asset

Sensor networks are comprised of several sensor nodes located around a structure. See Fig. 13.9. Each node becomes part of a multi-hop network, where data is relayed from one node to the base station by being passed between various other nodes. The exact path taken is flexible, dynamic, and depends on the network protocol and health of each node. Data paths can be optimized to use nodes with ample excess power, reducing errors by ensuring transmission success, or it can be optimized to use the least number of hops to reach the base station, and finally it can be optimized to utilize the least amount of energy. Furthermore, sensor nodes must be able to receive commands from the base station [31]. Attribute and location specific requests are common examples of commands sent to sensors nodes. For example, a user could request that all FBGs with strain readings above a certain threshold level report their location, strain, and temperature values. Alternatively, a user could request that all thermographic cameras in a geographical area send their latest image set.

Communication networks for TEI systems must be fault tolerant, meaning that if one sensor fails the entire network does not fail. This requires dynamic programming and protocols independent of the number of sensors and their location. Therefore, networks must also be scalable, and capable of handling numerous closely spaced sensor nodes [31]. Currently most communication networks are based on radio transmission of data, such as Bluetooth [30]. However, infrared and optical methods are also possible. That choice ultimately depends on the demands of the application and environment. Another major design consideration for wireless sensor networks is the necessary power supply.

Power Conservation, Supply, and Storage

While various sensors can be deployed in a TEI system to monitor a structure's behavior and performance, doing so requires power be available for every operation at every sensor. See Fig. 13.10. Sensors measure analog data, which is converted to a digital signal through an analog to digital convertor (ADC). This information is then sent to a microprocessor where any signal conditioning or calculations may be performed. A transceiver is used to send and receive signals from the base station. Onboard data storage may be available. A power supply is required to operate these components and operations.

Hardwired cables are not an option due to the required maintenance, potential for interference with the structure's operation, and the adverse impact on data gathering should a cable fail. Battery power is also an issue as batteries must be replaced at regular intervals. In small-scale projects these issues may not be of concern. However, large-scale projects with hundreds of sensors, battery replacement is not feasible. Therefore, sensors must be able to produce their own power. Considering this, energy harvesting, power storage, and conservation technologies have been the focus of much research over the past decade [33]. While solar, wind and tidal technologies are mature, they are more suited to large-scale applications. Most small-scale energy-harvesting research focuses on vibration based electromagnetic and piezoelectric harvesters, while thermoelectric generators (TEGs) and RF power transmitters have also been explored [33–36].

Converting mechanical strains into electrical energy through electromagnetic and piezoelectric methods is perhaps the most effective energy harvesting method. These devices can generate power from almost any type of vibration such as household appliances or bridge surfaces vibrating due to vehicular traffic [35, 36]. Electromagnetic generators are spring-mass-damper systems which vibrate a coil of wire inside a magnetic field thereby generating a current. This current may then be stored as a voltage in a capacitor or battery and used to power a sensor. A typical generator consists of a cantilevered arm holding magnets, which also perform the function of the massed damper [34]. See Fig. 13.11a, b.

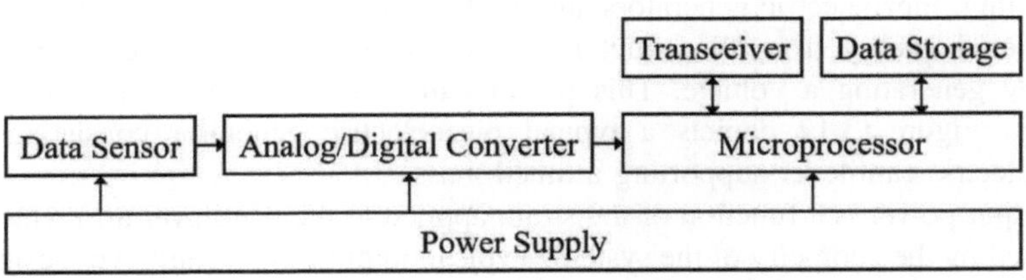

Fig. 13.10 Power supply needs at a single sensor node in a TEI system

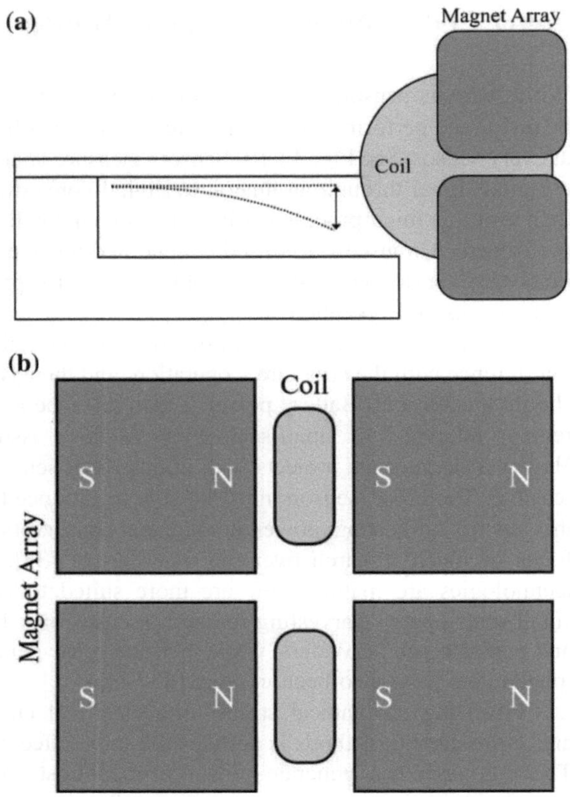

Fig. 13.11 **a** An electromagnetic energy harvester showing a coil of wire suspended from the end of the vibrating cantilever. **b** A magnet array surrounding the coil produces an electric current in the coil as it moves through the magnetic field

As the generator is vibrated by the structure, the cantilever starts to oscillate. The frequency of the oscillation is dependent on the input vibrations of the structure and can be tuned by the magnet mass and cantilever material properties. It is reasonable to expect around 50 μW of power to be generated from this type of electromagnetic energy harvester [33].

Piezoelectric-based generators are another popular area of research for energy harvesting. Piezoelectric generators and electromagnetic systems are similar in that vibrational energy from the structure is used to strain a piezoelectric material, thereby generating a voltage. This power can then be stored in a capacitor or battery. Figure 13.12 depicts a typical piezoelectric generator consisting of a piezoelectric cantilever supporting a tuned mass.

Output power is a function of the strain applied to the cantilever arm which can be tuned by the geometry of the system. Typical output from a cantilever design can range from 200 to 400 μW [35, 36]. Alternatively, piezoelectric patches can be applied directly to a vibrating surface. Such systems have the advantage of being sensitive to two-dimensional strain, but are more difficult to tune without the presence of the massed damper. A patch based piezoelectric system has been shown, theoretically, to be capable of generating up to 1 mW of power.

Fig. 13.12 A piezoelectric energy harvester

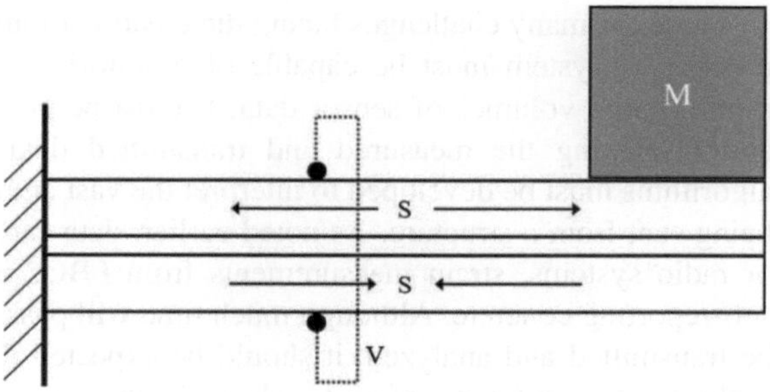

Because power is a precious commodity to TEI system based sensors, power conservation is an important aspect of practical applications [33]. Communication and computing tasks consume anywhere from 50 to 90% of power in most wireless sensors, and therefore require some form of power optimization and conservation. One form of power conservation is known as dynamic voltage scaling, where execution of sensor commands is slowed and operating voltages are reduced. In the first case, CPU processing is slowed to reduce total sensor idle time. Second, decreasing supply voltage when it is not needed can result in a cubic decrease in power consumption. Another power conservation strategy is known as Dynamic Power Management (DPM). In these systems components that are not in use are put into sleep mode, typically reducing power consumption by an order of magnitude. However, DPM-based systems often require the use of a control algorithm as transitioning in and out of sleep states can actually increase power consumption. Therefore, it is not ideal to rapidly transition a component into and out of its sleep state. DPM control algorithms attempt to calculate how long a particular component can be kept in its sleep state to achieve the most economical power consumption rate.

Infrastructure Management

Asset management is what turns mere structural health monitoring into a TEI system. Creating a well-integrated asset management system requires a thorough understanding of the critical structure parameters along with the sensor system outputs and responses to varying environmental conditions [37]. Asset managers must have an idea of the structure's current health and its expected deterioration rate. Furthermore, an integrated asset management system must allow decision makers the ability to identify the optimal response to available sensor data while weighing the associated required investment. Recall that the primary objectives of implementing a TEI system are to enhance public safety, prolong the useful life of an asset, and maintain or increase its value.

There are many challenges facing the creation of an asset management system or process. A system must be capable of monitoring, gathering, storing, and transmitting large volumes of sensor data. It must be possible to analyze, interpreting, and displaying the measured and transmitted data. Complex signal-processing algorithms must be developed to interpret the vast amount of information capable of being sent from a structure. As noted earlier, data can include images from thermal or radio systems, strain measurements from FBGs or cracking information from self-reporting concrete. Although much time will pass and vast amounts of data will be transmitted and analyzed, it should be expected that the sensor data will show nothing of significant interest, which is the desired circumstance. Thus, it is implausible to require consistent human monitoring of the reported data. Therefore, computer analysis techniques must reduce the sensor data to simple warning systems that can be pushed out to an asset manager, most likely on a geographical display on a cellular telephone or computer tablet [38–40]. Once a warning is displayed, the management team should be able to retrieve and review all relevant sensor data from to determine if there is a problem, the specifics of the problem, and the optimal resolution.

The multidimensional landscape of data complicates its review when evaluating an issue detected by an infrastructure sensor. For example, a highway bridge will undergo a visual inspection typically on an annual to biannual basis. Thermal imaging inspections may also be called for at some lesser interval. In an ongoing process, strain and temperature data may be collected with a FBG on a much shorter relative interval. This data will be collected throughout the calendar year with varying environmental conditions [41]. Asset managers must understand how sensor data can fluctuate over the course of the year due to temperature, humidity, and other environmental factors. In addition, sensor data is dispersed over a spatial dimension [42]. As an example, strain data may only be recorded at discrete points on a structure which are known to be indicators of potential structural failure. On the other hand, image data from thermal or radio sources can cover large areas of the structure. Asset managers must develop an understanding of how all this information fits together in order to make a well informed strategic decision regarding the optimal course of action. Clearly, the issue of asset management is complex on its own.

As mentioned previously, advanced signal-processing and decision-making algorithms can greatly improve the efficiency of asset management for TEI systems [42]. System information displays akin to modern electrical grid and traffic management systems should also be employed. However, interpretation of sensor data remains a major challenge to the creation of true TEI systems. The creation of standardized structure condition and damage indices can alleviate this issue. These indicators, based on the mechanical properties of the structure, must incorporate varying environmental conditions and lifetime of the structure, while adjusting for prior repairs and maintenance. Development of these indices must also be tested to ensure their correlation to structural failure. It has also been suggested that these indices should be based as close as possible to the raw recorded data, avoiding excessive modeling and assumptions. And finally, for robustness, several damage and condition indices should be used to improve the probability of detecting true structural failure.

Examples

Undoubtedly the most challenging aspect of creating a quality, reliable, and robust TEI system is to employ in the field the technologies available today and continue to develop all components of the system based on actual field data. As sensor, power and asset management technologies and practices evolve, implementation will become more routine and eventually common place. However, for now, best practices need to be developed, documented, and disseminated for the use of TEI systems in the field. Only through the actual use of these systems will it be possible to improve and advance each technology.

In general, there are two options for introducing TEI systems into a city's infrastructure. Preferably, monitoring systems can be installed during construction of a new infrastructure asset. It is anticipated that such an approach is possible with small-scale private projects where failure of any component in the TEI system would not become a structural, environmental, or legal issue. Attempting this approach in a large-scale public project, while certainly possible, would be expected in involve a great deal of effort to approve, implement, and activate the TEI system. The payoff from this approach is a fully integrated TEI system capable of producing data across the lifespan of the asset. Moreover, the safety, functionality, lifespan, and value of the asset should increase over time.

Alternatively, a TEI system may be attached or located remotely to an existing infrastructure asset. This approach could be valuable where preventative measures have been needed to maintain an asset. Use of a TEI system in this circumstance could again enhance the safety, functionality, lifespan, and value of the asset. Given that the TEI system would involve attaching sensors to a structure or collecting data without even contacting the asset, implementation of a system should be less problematic regardless of the nature of the project. Moreover, different systems could be easily employed to evaluate their functionality and value. In addition, the most appropriate sensor system can be applied to targeted areas of a structure, reducing installation costs and asset management complexities while at the same time delivering valuable information on the safety of the asset.

As part of the effort to disseminate best practices, below are two case studies of wireless sensor networks demonstrating different approaches to TEI implementation.

Preventative TEI System

Recent work at the second Jindo cable stayed bridge in South Korea is an excellent example of a large-scale wireless sensor network capable of capturing data for structural health monitoring [43, 44]. Originally built in 2006, the Jindo bridge is 12.5 m wide with a 344 m central span and two 70 m side spans. Prior to 2010, Jang et al. deployed a wireless sensor network on the Jindo bridge. A more traditional wired structural health monitoring system was already in place. However, the work by Jang et al. demonstrates some of the advantages and challenges associated with deploying sensor monitoring systems onto an existing structure.

As discussed earlier, several factors must be considered before a wireless sensor network is used as part of a structural monitoring system. In particular, asset managers must determine exactly what data they require, which commercial sensor(s) can provide the appropriate data, how the sensors will transmit the measured data, how the sensors will be powered, and finally how the data be will be analyzed, displayed, and interpreted. Because the Jindo bridge is subjected to several typhoons a year, wind induced vibrations are a cause for concern. Therefore, Jang et al. chose to use vibration and anemometer sensors. A total of 70 wireless sensor nodes were required due to the large span of the bridge.

Some of the sensors used on the Jindo Bridge were powered by single use batteries. While this is a readily available power source, single use batteries will require replacement or will define the life of a sensor. As an alternative, Jang et al. chose to power some of their sensors with rechargeable lithium-ion batteries connected to a solar cell. It was found that sensors using single use batteries had to be serviced every two months. In contrast, the solar powered sensors showed no appreciable decay in supply voltage over a 2-month period. However, the initial and recurring cost for the use of rechargeable batteries would clearly exceed that of the single use batteries. Thus, it will continue to be necessary to assess replacement/recharge intervals, first and recurring costs, and expected sensor life to determine which power supply is optimal for a given project.

The Jindo Bridge project also highlighted the impact that environmental conditions can have on a sensor. Base stations and sensors were enclosed in water-tight PVC boxes to protect them from wet and windy environmental conditions. Each base station consists of an industrial PC, an uninterrupted power supply and an internet modem. Enclosing the base state proved problematic as they overheated. Modifications were made to prevent further overheating events. However, the PVC encloses degraded the wireless signal transmission strength. Antennas had to be replaced to address this issue. Therefore, it is important to add time to a schedule and funds to a budget to address contingency issues.

Users of a TEI system will want to be familiar and comfortable with the software that operates the sensors and captures the measure data. The software associated with some sensors can be complex with a steep learning curve. The software used by Jang et al. is available at http://shm.cs.uiuc.edu/software.html. This software package has applications capable of sending sensor nodes to sleep to save power.

Another application can be used to instruct a handful of nodes to wakeup periodically to measure the environmental variables. If the measured data exceeds a certain threshold a signal is sent to the rest of the nodes, instructing them to wake up and record data. All of this, and several other applications, is controlled through an automatic monitoring application run from one of the two sensor base stations.

Vibration and wind data from the installed wireless sensor network was collected and used by Jang et al. as well as Cho et al. [44] to calculate the bridge's modal properties. A full analysis of the vibration data is presented in Cho et al. [44]. They made a comparison of the measured data to a finite element model. This is another example of the benefits of a sensor monitoring system. Measured data was not only useful in predicting and monitoring potential structural faults, it was also used to validate a computational model. The results of such an analysis can then be used to develop behavior prediction models used in the design of next generation infrastructure assets.

While the wireless sensor network on the Jindo Bridge is large, complex, and requires a large team of people to operate it, systems do not have to be so unwieldy. Users can consider small, agile, and targeted systems when the project conditions call for such a configuration.

Targeted TEI System

Hoult et al. [45, 46] have detailed their approach to upgrading traditional structures once visual inspections indicate the onset of degradation. The Ferriby Road Bridge near Hull, England is a variable width, skewed, three span structure nearly 45 m in length. Visual inspections showed minor cracking in the concrete deck slabs of approximately 0.1–0.2 mm in length. Tilting of the abutment bearings was also noted during the inspections. Hoult et al. used the opportunity to deploy a wireless sensor network to measure temperature, relative humidity, crack displacement (0.01 mm resolution), inclination (0.001° resolution), or both at the fault locations.

As with the Jindo Bridge, the Ferriby Bridge wireless sensor network project highlighted some of the advantages and challenges associated with deploying a TEI system. Hoult et al. were able to install a system within two days while minimizing traffic delays. This was largely due to the wireless nature of the monitoring system. A wired system would require significantly more installation time. Hoult et al. found further advantage in the ability to place each sensor node in a targeted location. This resulted in highly efficient use of the sensors which collected and transmitted only relevant data.

Hoult et al. do note some challenges to deploying a TEI system. Supplying power to the sensor, data acquisition, and data storage systems is a primary challenge for a wireless sensor network. Hoult et al. relied on lithium-ion batteries for each sensor node with calculated lifespans of 1.5–4 years. However, the data storage and server system required a more robust solution. Here, a 12 V battery, charged by a 64 W solar panel was used. Due to the remoteness of the bridge location, accessing the data server required the use of a mobile phone internet connection. Finally, wireless connectivity issues were encountered. This required the replacement of some sensor node antennas to rectify the issue. Again, it is important to add time and funding to a project to plan for contingencies.

The wireless sensor network installed on the Ferriby Road Bridge provided useful short-term and long-term data. Humidity and temperature readings were taken at each sensor node, in addition to data server location, allowing for data corrections in response to environmental changes. Crack widths were monitored and showed no significant changes over a 5-month period, indicating that crack growth on the bridge is occurring on a yearly timeframe as opposed to monthly. This was useful information for asset managers, because it allows them to plan for repairs well ahead of time. Furthermore, bearing pad inclination was shown to be correlated to measured temperature over both short- and long-term observations. Short-term variations were shown to be due to expansion and contraction of the bearing, while long-term variations are attributed to the expansion of the bridge deck itself. This demonstrates that asset managers must have knowledge of the overall context of the measured data, otherwise they run the risk of making poor decisions based on an incorrect interpretation of the data.

Conclusion

As sensor technology, power supplies, communication networks, data storage, and data interpretation methodologies evolve and user experience expands, more cities will choose to invest in Technology-Enhanced Infrastructure (TEI) systems. Cities will use these systems to monitor their assets to ensure public safety and maintain asset value. Transportation, health, education, water treatment, water resource, storm water management, waste management systems, and more can all benefit from the implementation of a TEI system.

Even though TEI systems continue to develop, it is important to consider what is next for monitored infrastructure. Yet to be considered by many is the use of TEI to develop predictive models for approximating the useful life of an asset. TEI networks capable of predicting the remaining life of a structure would be an invaluable resource for infrastructure asset managers. Having knowledge of an approximate time when major and minor repairs would be required for a structure would allow asset managers to schedule and budget these repairs in a more efficient and cost effective manner. Being able to purchase repair materials in advance when prices might be lower, scheduling manpower to limit idle time, and mobilizing manpower in advance of a repair would ultimately improve the economic efficiency of a city's entire infrastructure inventory. To realize these advantages requires an empirical lifetime prediction model based upon actual field data captured by smart sensors. Not only will such a model contribute to more resilient structures it will also extend the functionality and economic feasibility of monitoring systems.

Another future need will be monitoring systems capable of measuring properties not previously achieved by other devices, in a manner not previously conceived, or both. A research collaboration housed in the School of Engineering at Texas State University is employing principles from physics, electrical engineering, and construction materials to develop a novel approach for the measurement of stress in concrete. This approach employs radio frequency illumination to measure stress. In this study, properties of concrete specimens are measured remotely using lensed horn antennas connected to a vector network analyzer. The parameters being measured are related to concrete strength and are known to be correlated to the age of concrete. It will be possible to employ such a monitoring system to establish an empirical lifetime prediction model. Once a predictive model has been validated via use of the proposed monitoring system, research will examine the sensitivity of the model to variations in the concrete's constituent materials and proportions. This is just one example of a novel monitoring sensor. Only time will tell what other types of sensors and TEI systems will be developed over the next decade and more.

References

1. Adibhatla V, Henke G, Atwater P (2015) Street quality identification device. Paper presented at Bloomberg data for good exchange conference, New York City, 28 Sept 2015
2. Ford Motor Company (2016) 2017 Fusion sport. http://www.ford.com/cars/fusion/trim/sport/ Accessed Sept 16 2016
3. Moyo P, Brownjohn JMW, Suresh R et al (2005) Development of fiber Bragg grating sensors for monitoring civil infrastructure. Eng Struct 27:1828–1834
4. Majumder M, Gangopadhyay TK, Chakrabortty AK et al (2008) Fiber Bragg gratings in structural health monitoring: present status and applications. Sens Actuator 147:150–164
5. Casas JR, Cruz PJS (2003) Fiber optic sensors for bridge monitoring. J Bridge Eng 8(6):262–373
6. Chan THT, Yu L, Tam HY et al (2006) Fiber Bragg grating sensors for structural health monitoring of Tsing Ma bridge: background and experimental observation. Eng Struct 28:648–659
7. Schulz WL, Conte JP, Udd E et al (2000) Static and dynamic testing of bridges and highways using long-gage fiber Bragg grating based strain sensors. In: Environmental and industrial sensing. International society for optics and photonics, vol 4202, pp 79–86
8. Li HN, Li DS, Song GB (2004) Recent applications of fiber optic sensors to health monitoring in civil engineering. Eng Struct 26:1647–1657
9. D'Alessandro A, Ubertini F, Materazzi AL (2016) Self-sensing concrete nano composites for smart structures. Int J Civil Environ Struct Constr Archit Eng 10(5):576–581
10. Chen PW, Chung DDL (1993) Carbon fiber reinforced concrete for smart structures capable of non-destructive flaw detection. Smart Mater Struct 2:22–30
11. Fu X, Chung DDL (1996) Self-monitoring of fatigue damage in carbon fiber reinforced cement. Cem Concr Res 29:435–439
12. Chung DDL (1998) Self-monitoring structural materials. Mater Sci Eng R22:57–78
13. Shi ZQ, Chung DDL (1999) Carbon fiber-reinforced concrete for traffic monitoring and weighing in motion. Cem Concr Res 29:435–439
14. Büyüköztürk O (1998) Imaging of concrete structures. NDT&E Int 31(4):233–243
15. Arunachalam K, Melapudi VR, Upda L et al (2006) Microwave NDT of cement-based materials using far-field reflection coefficients. NDT&E Int 39:585–593

16. Jamil M, Hassan MK, Al-Mattarneh HMA et al (2013) Concrete dielectric properties investigation using microwave nondestructive techniques. Mater Struct 46:77–87
17. Bungey JH (2004) Sub-surface radar testing of concrete: a review. Constr Build Mater 18:1–8
18. Lin J, Liu CR (2004) Measurement of concrete highway rough surface parameters by an X-band scatterometer. IEEE Trans Geosci Remote 42(6):1188–1196
19. Oyama Y, Zhen L, Tanabe T et al (2009) Sub-terahertz imaging of defects in building blocks. NDT&E Int 42:28–33
20. Büyüköztürk O, Yu TY (2009) Far field radar NDT technique for detecting GFRP debonding from concrete. Constr Build Mater 23:1678–1689
21. Sakagami T, Kubo S (2002) Development of a new non-destructive testing technique for quantitative evaluations of delimitation defects in concrete structures based on phase delay measurement using lock-in thermography. Infrared Phys Technol 43:311–316
22. Titman DJ (2001) Applications of thermography in non-destructive testing of structures. NDT&E Int 34:149–154
23. Clark MR, McCann DM, Forde MC (2003) Application of infrared thermography to the non-destructive testing of concrete and masonry bridges. NDT&E Int 39:265–275
24. Maierhofer C, Arndt R, Röllig M (2007) Influence of concrete properties on the detection of voids with impulse-thermography. Infrared Phys Technol 49:213–217
25. Maierhofer C, Arndt R, Rieck C et al (2006) Application of impulse-thermography for non-destructive assessment of concrete structures. Cem Concr Compos 28:393–401
26. Washer G (1998) Developments for the non-destructive evaluation of highway bridges in the USA. NDT&E Int 34:245–249
27. Duffy GS, Farina SB (2009) Development of an embeddable sensor to monitor the corrosion process of new and existing reinforced concrete structures. Constr Build Mater 23:2746–2751
28. McCarter WJ, Chrisp TM, Butler A et al (2001) Near-surface sensors for condition monitoring of cover-zone concrete. Constr Build Mater 15:115–124
29. McCarter WJ, Starrs G, Chrisp TM (2000) Electrical conductivity, diffusion, and permeability of Portland cement-based mortars. Cem Concr Res 30:1395–1400
30. Spencer BF Jr, Ruiz-Sandoval ME, Kurata N (2004) Smart sensing technology: opportunities and challenges. Struct Control Health Monitor 11:349–386
31. Akyildiz IF, Su W, Sankarasubramaniam E et al (2002) A survey on sensor networks. IEEE Commun Mag 40:102–114
32. Hsieh YM, Hung YC (2009) A scalable IT infrastructure for automated monitoring systems based on the distributed computing technique using simple object access protocol web-services. Automat Constr 18:424–433
33. Park G, Rosing T, Todd MD et al (2008) Energy harvesting for structural health monitoring sensor networks. J Infrastruct Syst 14(1):64–79
34. Beefy SP, Torah RN, Tudor MJ et al (2007) A micro electromagnetic generator for vibration energy harvesting. J Micromec Microeng 17:1257–1265
35. Roundy S, Wright PK (2004) A piezoelectric vibration based generator for wireless electronics. Smart Mater Struct 13:1131–1142
36. Erturk A (2011) Piezoelectric energy harvesting for civil infrastructure system applications: Moving loads and surface strain fluctuations. J Intell Mater Syst Struct 22(17):1959–1973
37. Catbas FN, Aktan AE (2002) Condition and damage assessment: issues and some promising indices. J Struct Eng 128(8):1026–1036
38. Malekzabeh M, Atia G, Catbas FN (2015) Performance-based structural health monitoring through an innovative hybrid data interpretation framework. J Civ Struct Health Monitor 5:287–305
39. Amini F, Karami K (2012) Damage detection algorithm based on identified system Markov parameters (DDA/ISMP) in building structures with limited sensors. Smart Mater Struct 21 (5):055010
40. Hsieh YM, Hung YC (2009) A scalable IT infrastructure for automated monitoring systems based on distributed computing technique using simple object access protocol Web-services. Automat Constr 18:424–433

41. Aktan AE, Catbas FN, Grimmelsman KA et al (2000) Issues in infrastructure health monitoring for management. J Eng Mech-ASCE 126(7):711–724
42. Li H, Ou J (2016) The state of the art in structural health monitoring of cable-stayed bridges. J Civ Struct Health Monitor 6:43–67
43. Jang S, Jo H, Cho S et al (2010) Structural health monitoring of a cable-stayed bridge using smart sensor technology: deployment and evaluation. Smart Struct Syst 6(5–6):439–459
44. Cho S, Jo H, Jang S et al (2010) Structural health monitoring of a cable-stayed bridge using smart sensor technology: data analysis. Smart Struct Syst 6(5–6):461–480
45. Hoult NA, Fidler PRA, Hill PG et al (2010) Long-term wireless structural health monitoring of the Ferriby Road Bridge. J Bridge Eng 15:153–159
46. Hoult NA, Fidler PRA, Hill PG et al (2008) Wireless structural health monitoring at the Humber Bridge UK. In: Proceedings of the institution of civil engineers-bridge engineering, vol 161 no 4, pp 189–195

Index

S. McClellan et al. (eds.), *Smart Cities*, DOI 10.1007/978-3-319-59381-4

MIX
Papier aus verantwortungsvollen Quellen
Paper from responsible sources
FSC® C105338

If you have any concerns about our products,
you can contact us on
ProductSafety@springernature.com

In case Publisher is established outside the EU,
the EU authorized representative is:
Springer Nature Customer Service Center GmbH
Europaplatz 3, 69115 Heidelberg, Germany

Printed by Libri Plureos GmbH
in Hamburg, Germany